*access to geography*

# GLACIAL *and* PERIGLACIAL ENVIRONMENTS

*David Anderson*

I am grateful to Adrian Parker for many years of collaboration on fieldwork in Scotland and England, beginning during our days as research students at the SoGE, Oxford, which has inspired several of the themes treated in this book. I particularly thank Emma, my wife and colleague, for her advice and support while writing it.

**Acknowledgements**
The publishers would like to thank the following individuals, institutions and companies for permission to reproduce copyright illustrations in this book:

Professor C Ballantyne, University of St Andrews, page 103 (below); CK Ballantyne and C Harris, *The Periglaciation of Great Britain*, 1994, Cambridge University Press, page 96; *The Scottish Geographical Journal*, Volumes 116 (no. 2) and 117 (no. 4), Edinburgh University Press, pages 67, 103 (above); Douglas I Benn and David JA Evans, *Glaciers and Glaciation*, 1998, Edward Arnold © 1998 Douglas I Benn and David JA Evans. Reprinted by permission of Hodder Arnold, page 51; Geoscience Features Picture Library, page 2; D Duff and PMD Duff, *'Holmes' Principle of Physical Geology*, Nelson Thornes Limited, page 33; A Clowes and P Comfort, *Process and Landform; An Outline of Contemporary Geomorphology 1 Edn*, Pearson Education Limited, pages 62, 79, 99; I Fishpool, *Glaciation and Deglaciation, Geography Review*, Volume 9, Issue 3, 1996, Philip Allan Updates, pages 70, 71; Taylor and Francis (Routledge), pages 2, 81.



Note about the Internet links in the book. The user should be aware that URLs or web addresses change regularly. Every effort has been made to ensure the accuracy of the URLs provided in this book on going to press. It is inevitable, however, that some will change. It is sometimes possible to find a relocated web page, by just typing in the address of the home page for a website in the URL window of your browser.

Orders: please contact Bookpoint Ltd, 130 Milton Park, Abingdon, Oxon OX14 4SB. Telephone: (44) 01235 827720. Fax: (44) 01235 400454. Lines are open from 9.00 to 5.00, Monday to Saturday, with a 24-hour message answering service. You can also order through our website www.hoddereducation.co.uk.


First Published 2004 by
Hodder Education,
An Hachette UK Company,
338 Euston Road, London NW1 3BH

Impression number 10
Year 2014



Cover photo: The Teton Range (Wyoming, USA) viewed from the Snake River Overlook, © David Anderson.
Produced by Gray Publishing, Tunbridge Wells, Kent
Printed and bound by CPI Group (UK) Ltd, Croydon CR0 4YY
A catalogue record for this title is available from the British Library

ISBN 978 0340 81247 1

# Contents

# 1 Present and Past Distribution of Glaciers and Ice Sheets

**KEY WORDS**

**Glacial**: a phase during an ice age when ice sheets extend from the polar regions into the mid-latitudes, and global climate becomes colder and drier.
**Glacier**: a mass of land ice that deforms under its own weight and flows downslope.
**Holocene**: the second epoch of the Quaternary, and the current interglacial phase that began about 11 500 years ago.
**Ice sheet**: a continental-sized glacier, at least 50 000 km$^2$ in area, that is dome-shaped with flow of ice outward from the centre.
**Interglacial**: a phase of relative warmth between glacial phases when ice sheets are greatly reduced in size and restricted to the polar regions.
**Pleistocene**: the first epoch of the Quaternary, consisting of all the glacial/interglacial cycles preceding our present interglacial phase.
**Quaternary**: the most recent period of geological time, preceded by the Tertiary Period, and divided into two epochs.

## 1 Types of Glaciers

Glaciers are generally classified by size. The smallest glaciers, only just large enough to be glaciers rather than just stationary snowfields, are known as **cirque glaciers** (also termed **corrie glaciers**). These glaciers usually cover an area of between 0.5 and 10 km$^2$. These small glaciers form in shady, sheltered depressions high up on the sides of mountains where the snow accumulation is greatest, and they can remain confined to these depressions, scouring and deepening them over time (as explained in Chapter 3) into armchair-shaped hollows called cirques (or corries). If a glacier grows too large to be contained in its cirque and extends down into a valley between mountains, it can be termed a **valley glacier**. Valley glaciers vary greatly in size, covering tens to thousands of square kilometres, depending upon the width and length of the valley that they fill. Smaller **tributary glaciers** often feed larger valley glaciers. These tributary glaciers occupy their own smaller valleys that join a main valley. The Aletsch Glacier in southern Switzerland, the largest glacier in Europe, is a valley glacier 23.6 km in length that covers approximately 120 km$^2$.

The Aletsch Valley Glacier, Switzerland

With further expansion, several valley glaciers can merge to create a **highland ice field** in which a large percentage of an upland region is covered by glacier ice, with the exception of the tallest peaks (termed **nunataks**) that protrude above the ice (Figure 1). Highland ice fields overlap in areal coverage with large valley glaciers, but the largest highland ice fields can cover many thousands of square kilometres. Although relatively small today compared to its size during the last glacial phase, the Columbia ice field in Canada (located at 52°N, 117°W) is the largest body of ice in the Rocky Mountains, covering about 325 km$^2$. A number of different **outlet glaciers** move outwards from the ice field. A large glacier that extends from a highland area into the foothills and surrounding lowlands can be termed a **piedmont glacier**. For example, the Wilson Piedmont Glacier in Antarctica, due south of New Zealand, extends down from the Transantarctic Mountains and across the coastal plain along the western shore of the Ross Sea. It is over 50 km wide, and is fed by many smaller valley glaciers coming out of the mountains.

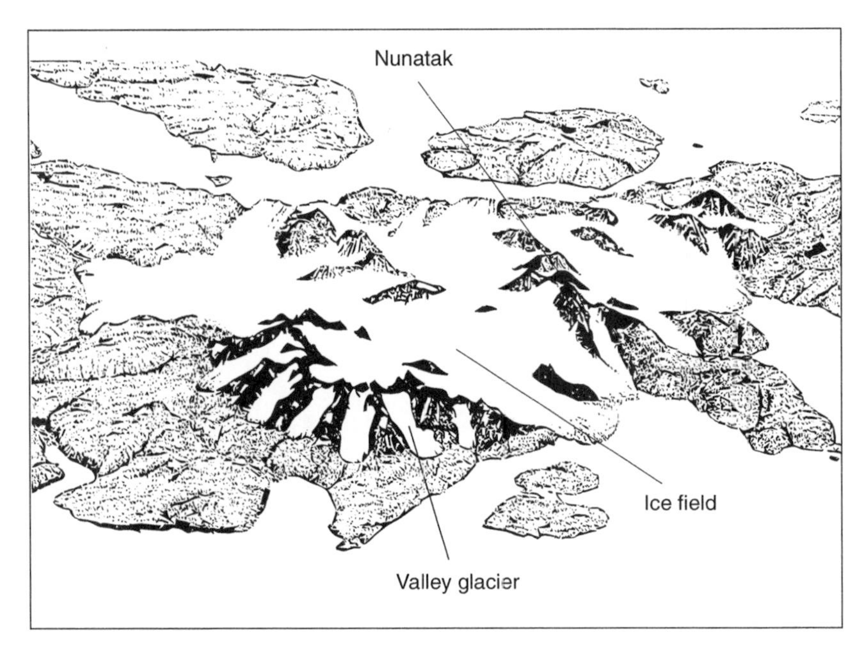


**Figure 1** A highland ice field with protruding nunataks (Cuillin Hills, Isle of Skye during the Loch Lomond Stadial, from Ballantyne, 1989)

At the largest scale, the extension of a highland ice field beyond the confines of the mountains to cover tens of thousands, or even millions, of square kilometres of lowland constitutes an **ice cap** or **ice sheet**. The two terms are often used interchangeably, although some glaciologists prefer to use 'ice sheet' to represent continental-sized accumulations of glacier ice (at least 50 000 km$^2$ in area), such as the Greenland or Antarctic ice sheets.

# 2 The Present-day Distribution of Glaciers

## a) Modern glacier ice cover

Today, glaciers cover approximately 10% (about 15.8 million km$^2$) of the Earth's land area, with about 85% of all glacier ice being contained within Antarctica, shared between the west and east Antarctic ice sheets. The second largest accumulation of glacier ice is the Greenland ice sheet, covering about 1.8 million km$^2$ and making up approximately 11% of the Earth's total ice cover. The remaining glacier cover is distributed among ice caps, highland ice fields and smaller glaciers in mountains and high-latitude regions around the globe (Table 1).

**Table 1** Estimates of the area and volume of present-day regions of glacier ice. *Source*: Smithson *et al.* (2002)

| Region | Est. area ($10^6$ km$^2$) | Est. volume ($10^6$ km$^3$) |
|---|---|---|
| Antarctica | 13.50 | 32.0 |
| Greenland | 1.80 | 2.6 |
| Arctic basin | 0.24 | – |
| Alaska | 0.05 | – |
| Rest of N. America | 0.03 | – |
| Andes | 0.03 | – |
| European Alps | 0.004 | – |
| Scandinavia | 0.004 | – |
| Asia | 0.12 | – |
| Africa | 0.0001 | – |
| Australasia | 0.001 | – |
| **Total** | **15.8** | **35.0** |

Outside of Antarctica and Greenland, parts of Canada fringing the Arctic Ocean (such as Baffin, Devon and Ellesmere Islands) and the Pacific coast of Alaska contain the most glacier ice. There are also significant accumulations in some of the world's great mountain ranges, such as the Himalayas, the Andes, the Rockies and Cascades of North America, and the European Alps (Figure 2). Of all the freshwater in the world, approximately 75% is locked up in ice, and glaciers contain about 1.8% of all the water on Earth.

## b) Factors influencing the distribution of glaciers

The two most important factors determining the present-day distribution of glaciers are latitude and altitude. Owing to the lower angle at which the sun's rays hit the ground at higher latitudes, much less solar energy is received per unit area near the poles than near the equator – this is the reason why average annual temperatures increase toward the equator and decrease toward the poles.

Because the equatorial regions receive the most solar energy per unit area, temperatures at or near sea level are far too high for snowfall and snow accumulation. However, glaciers are found in some equatorial regions, such as in central Africa, Indonesia and Ecuador, but only at altitudes in excess of 4000 m where mean annual temperatures are low because of the high elevation. As altitude increases, lower atmospheric pressure causes air to expand its volume, lose energy and decrease in temperature. Reduced temperature due to this process is known as an **adiabatic** temperature decline. Dry air

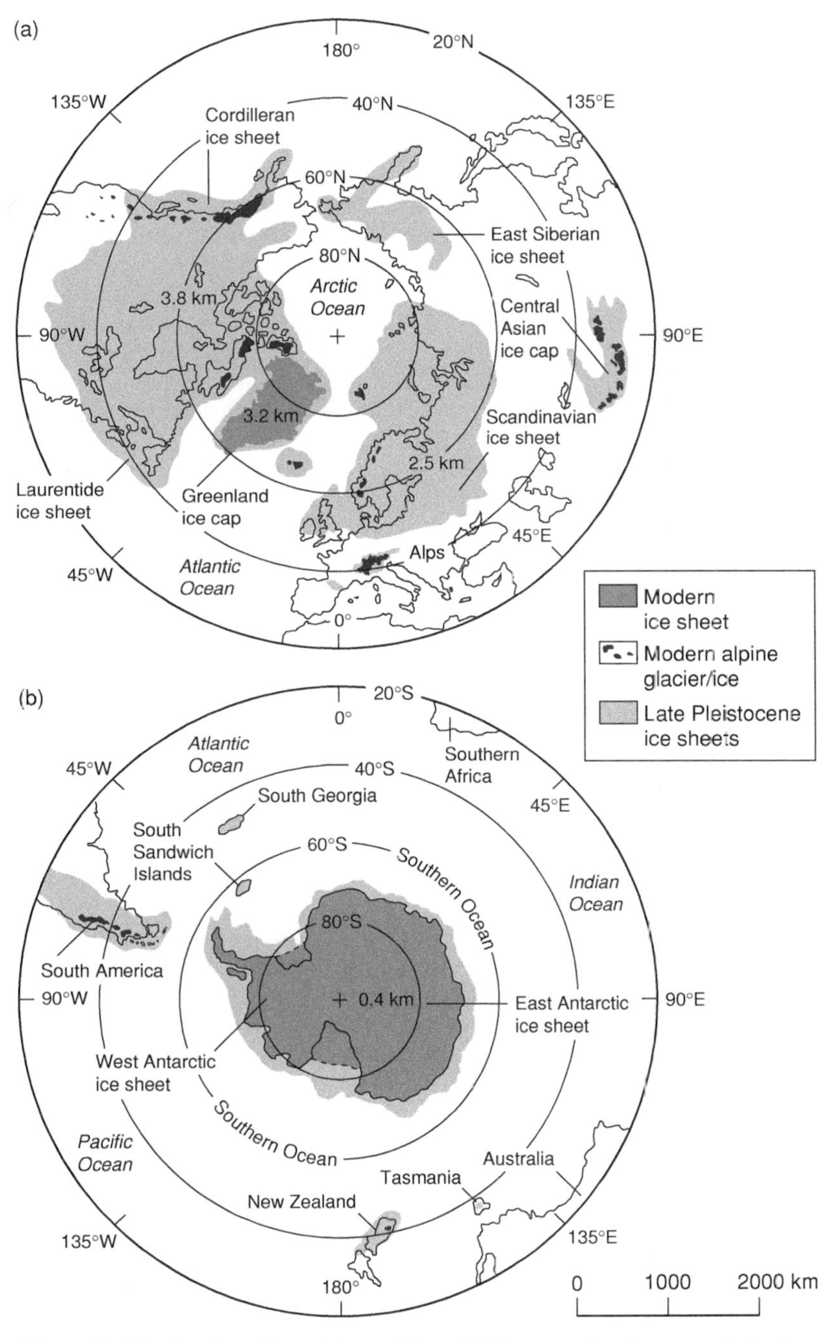


**Figure 2** Distribution of modern and Late Pleistocene ice sheets and glaciers in the Northern and Southern Hemispheres. *Source*: Briggs *et al.* (1997).

declines in temperature by 1°C for every 100 m gain in altitude (the dry adiabatic lapse rate). If the air is humid, the temperature decline is slowed to between 0.4 and 0.9°C per 100 m (the saturated adiabatic lapse rate) depending upon the amount of moisture condensing out of the air.

The city of Dar es Salaam in Tanzania, Africa (6°50′S, 39°18′E), for example, is at an altitude of only 14 m above sea level and has an average daily maximum temperature of 29.6°C throughout the year. Taking the saturated adiabatic lapse rate (averaging 0.6°C per 100 m), an altitude gain of 5000 m above Dar es Salaam is necessary before daily maximum temperatures will be around freezing point, making it possible for snow to survive and accumulate over long periods of time. This is achieved at Mount Kilimanjaro in Tanzania (to the north-east of Dar es Salaam and 5892 m high) where there are permanent snowfields and small glaciers.

Moving to the mid-latitudes, mean annual temperatures at sea level are less than at the equator, and therefore average temperatures of around 0°C occur at lower altitudes. At about 45°N latitude in the Rocky Mountains of the USA glaciers can be found at around 3000 m above sea level. Above 60°N, many glaciers in Norway are found between about 1200 and 2000 m altitude, and within the Arctic Circle, for example in Svalbard (between 77 and 80°N) glaciers extend down to sea level. However, the relationship between the latitude and the altitude of glacier formation is not completely consistent because of the influence of other factors on snow accumulation. For example, glaciers in the Cascade Mountains close to the Pacific Ocean form at lower altitudes (often more than 1000 m lower) than more inland glaciers of the Rockies located along the same line of latitude because the Cascades have higher rates of snowfall.

On a more local scale, both slope aspect and relief affect the distribution of glaciers. North-facing slopes are shadier in the Northern Hemisphere and therefore more conducive to snow accumulation. With westerly winds dominating in the mid-latitudes, snow is more sheltered on the eastern side of mountains. Acting together, these two factors often cause the greatest snow accumulation, and hence lowest altitude glaciers, to occur on the north-eastern flanks of mountains. In most Southern Hemisphere mountain ranges, such as the Southern Alps of New Zealand and the Chilean Andes, prevailing winds are also westerly causing the south-eastern flanks of mountains to have the lowest altitude glaciers. The 'all-sided glaciation level', meaning the altitude at which glaciers can form on all aspects of a mountain, can be up to 300 m higher than the more localised minimum glacier level. Relief is also important in determining the surface upon which glaciers accumulate. In very steep, high relief mountains, glacier ice tends to be less abundant because there is not very much low-angle ground at high altitudes to hold the snow and allow it to accumulate.

# 3 Ice Ages Through Geological Time

## a) The 'Ice Age' discovery

The 19th century discovery that the world has experienced an 'Ice Age' was of great importance for our understanding of the Earth and its history. It was the Swiss scientist Louis Agassiz in the late 1830s who introduced the concept of continental glaciation, in which ice sheets extended over large areas of the mid-latitudes. His theory of 'a great Ice Age' made it possible to explain for the first time many of the landforms of glaciation that are discussed later in this book. The first scientific observations of existing glaciers were made in the European Alps, paving the way for identifying signs of glaciation in places where glaciers no longer exist today, such as in the British Isles. Since the idea of an 'Ice Age' was proposed, evidence for past glaciation has been found over large areas of North America and Eurasia from high to mid-latitudes (southwards to approximately 40°N).

In addition to providing the key to explaining landscapes across much of the globe (particularly at high latitudes and altitudes), the discovery of an 'Ice Age' also revolutionised our understanding of the Earth's climate and how it changes over long periods of time. Research in the 20th century, especially since the 1970s, has shown that the planet has experienced not one 'Ice Age' as originally conceptualised in the 19th century, but several ice ages; and we currently occupy the most recent ice age – the **Quaternary** Ice Age. Ice ages themselves are subdivided into phases of extensive glaciation (**glacials**) separated by phases of warmer conditions with retreat of glaciers (**interglacials**). As explained later in this chapter, the large changes in climate that occurred between, and during, glacials are now known to have been remarkably rapid, rather than slow and gradual as once thought.

## b) The Quaternary

We are currently living during the period of geological time known as the **Quaternary**. It began around 2 million years ago as the Tertiary Period ended, and it is distinguished from the preceding Tertiary by the onset of global cooling, which led to the expansion of ice sheets across large areas of the globe at high to mid-latitudes. For this reason the Quaternary Period is also known as the Quaternary Ice Age. The exact time when it became cold enough to mark the transition between the Tertiary and the Quaternary is a matter of interpretation still debated among geologists, and estimates range between 2.6 and 1.8 million years ago.

Although glacier ice is not as extensive today as it has been at other times during the Quaternary, the ice age continues. This is because the existence of the Antarctic and Greenland ice sheets (and smaller glaciers elsewhere) sets our time apart from most other periods of Earth's history when there was little or no permanent ice anywhere on the globe. For example, during the Cretaceous Period from 135 to 65 million years ago (dinosaurs became extinct at the end of the Cretaceous), average global temperatures were over 5°C warmer than today and the Earth was ice-free. Sea level was much higher than at present, and there were forests at high latitudes. Prior to the Quaternary Period, the last time the Earth experienced an ice age was around 280 million years ago during the Carboniferous and Permian periods. Throughout the whole of Earth's history, there is geological evidence for at least seven alternations between ice ages and warmer periods.

## 4 Multiple Glacials and Interglacials During the Pleistocene

### a) Characteristics of the Pleistocene

The Quaternary Period is subdivided into two epochs of geological time – the **Pleistocene** and the **Holocene**. The Pleistocene Epoch covers the time span from the beginning of the Quaternary to about 11 500 years ago, when the most recent glacial ended and the present interglacial began. The Holocene interglacial in which we now live is similar to previous interglacials, and it can be argued that there is no reason to mark the present interglacial as the start of a new geological epoch. However, what distinguishes our interglacial from previous ones is that it has seen the development of agriculture and the growth of civilisation.

When viewed in relation to the very long timescale of Earth's geological history, the relatively cold Quaternary Period can be regarded as a single 'ice age', which continues to this day (as described in the previous section). However, during the Quaternary Ice Age itself, conditions have not been uniformly cold. Instead, there has been much variation in the Earth's climate over shorter timescales, and large ice sheets have advanced and retreated many times as the climate has repeatedly shifted between colder and warmer, causing the oscillation between glacials and interglacials. As the Holocene refers only to our present interglacial, the multiple phases of glaciation (and intervening interglacials) that characterise the Quaternary Ice Age are all contained within the time-span of the Pleistocene.

Glacial phases during the Pleistocene have left evidence on land in the form of erosional and depositional features created by glaciers (described in Chapters 3 and 4). However, the landforms produced by glaciers during earlier glaciations have usually been reshaped or destroyed by later glaciations, making reconstruction of the pattern of past glaciation difficult. Only the four most recent Pleistocene glacial phases are easily recognised from evidence on land. The names given to these four glacials are different in different parts of the world. Starting with the most recent and moving back in time, in the British Isles they are known as the **Devensian**, the **Wolstonian**, the **Anglian** and the **Beestonian** glaciations. In the European Alps the corresponding names are the Würm, Riss, Mindel and Günz glaciations, and in North America the Wisconsin, Illinoian, Kansan and Nebraskan glaciations.

In comparison with the evidence on land, evidence from the sea floor provides a much longer and more complete record of past glacial/interglacial cycles. This is because in oceans the shells and skeletons of marine organisms are continually deposited on the sea bed, building up undisturbed layers of sediment on the ocean floor over many thousands, and in some places even millions, of years. By sampling this ocean sediment and analysing the chemistry of shells dating from different time periods it is possible to infer what the ocean environment was like (for example, sea temperatures) at the times when the marine organisms were alive. Continuous samples taken down through ocean sediment (sediment cores) in many of the world's oceans have made it possible to produce detailed reconstructions of how the ocean environment has changed during the Pleistocene that have been important for understanding the pattern of glaciation on land.

This has been achieved mainly by analysing the oxygen isotope ratios (the ratio of $^{18}O/^{16}O$) found within the shells of marine organisms that have been deposited on the ocean floor.

- More of the heavier type of oxygen ($^{18}O$ that has ten neutrons as opposed to the lighter $^{16}O$ that has just eight) within shells indicates times when more of the Earth's water was locked up in ice sheets and glaciers.
- Shells with less $^{18}O$ relative to $^{16}O$ were formed during times when ice sheets and glaciers were less extensive.
- This is because when water is evaporated from the oceans, $H_2O$ containing $^{16}O$ is more easily evaporated than $H_2O$ containing $^{18}O$.
- In order to build up large ice sheets, a lot of water from the oceans must be evaporated and deposited as snow over landmasses. This process results in taking more $H_2{}^{16}O$ out of the oceans while leaving $H_2{}^{18}O$ behind, causing an enrichment in $^{18}O$ of sea water during glacials.
- As marine organisms build their shells with oxygen from surrounding sea water, the increased $^{18}O/^{16}O$ ratio of the sea water during glacials becomes preserved in the shells of the organisms that were living at the time. The $^{18}O/^{16}O$ ratios reach their maximum when global ice volume is at a maximum.

Depending on how the oxygen isotope ratios in ocean sediments are interpreted, there appears to have been between 30 and 50

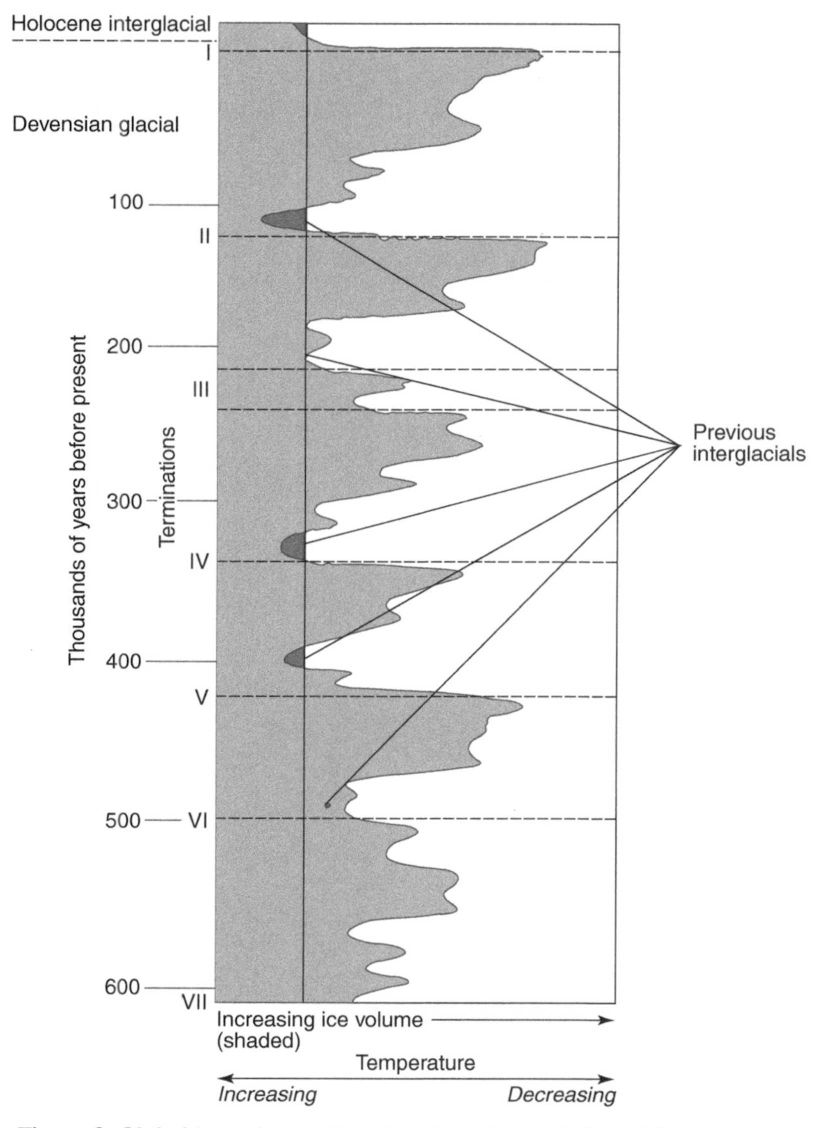


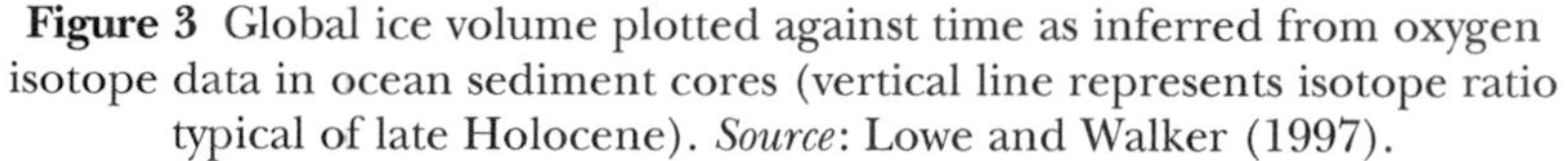
**Figure 3** Global ice volume plotted against time as inferred from oxygen isotope data in ocean sediment cores (vertical line represents isotope ratio typical of late Holocene). *Source*: Lowe and Walker (1997).

cold/temperate cycles since the Pleistocene began. During the earlier part of the Pleistocene, shifts between glacial and interglacial phases occurred approximately every 41 000 years, but after about 900 000 years ago the oscillations became longer, showing a periodicity of around

100 000 years. For example, over the past 600 000 years there have been six pronounced phases of glaciation, each followed by an interglacial when the global ice volume decreased substantially (Figure 3). The pattern also shows that the global ice volume tends to build up relatively gradually during a glacial, but decreases rapidly at the beginning of an interglacial.

The majority of the Pleistocene has been characterised by colder conditions with a greater global ice volume than exists today, and times of interglacial warmth similar to the present Holocene are relatively short. Figure 3 shows that over the past 600 000 years there have been only three short periods of time when temperatures were as warm as they are today. Over the past 1 million years or so, today's type of climate has only occurred for around 10% of the time, whereas most of the time it has been colder, and ice sheets and glaciers have been much more extensive.

## b) Cause of glacial/interglacial cycles during the Pleistocene

Long-term changes in the Earth's orbit around the sun are the main cause for the oscillations between glacial and interglacial conditions that have occurred during the Pleistocene. The idea that changes in the Earth's orbit could explain the expansion and retreat of continental ice sheets dates back to the 19th century when the Scottish scientist James Croll first speculated about how such changes might be linked to climate and to glaciation. However, it was the Serbian Milutin Milankovitch, in the early 20th century, who first calculated how the amount of solar energy received by the Northern and Southern Hemispheres in different seasons changes as the Earth's orbit changes.

The **Milankovitch astronomical theory** of glacial periods takes into account three characteristics of the Earth's orbit. First, the orbit changes from being more elliptical to more circular and back again over a period of about 96 000 years. This cycle is known as the **eccentricity** of the orbit. Second, the tilt of the Earth's axis varies from 21.8° to 24.4° relative to the plane of its orbit (the plane of the ecliptic) over a period of 42 000 years. This is known as the **obliquity of the ecliptic**, and today the Earth's axis is tilted at 23.5°. Third, the Earth wobbles on its axis like a top causing long-term changes in where different seasons occur along the Earth's orbital path. For instance, today the Northern Hemisphere summer occurs when the Earth is near its furthest point from the Sun on its orbit (aphelion), whereas around 10 000 years ago the Northern Hemisphere summer occurred when the Earth was nearest the Sun on its orbit (perihelion). This cycle, known as the **precession of the equinoxes**, averages out over a period of 21 000 years.

The Milankovitch theory combines all three of these cycles to determine when solar energy is minimised or maximised in the

Northern Hemisphere summer in explaining when glacials and interglacials occur. The input of solar energy into the Northern Hemisphere is more important than the Southern Hemisphere for controlling glaciation during the Quaternary Period because there is much more land at high latitudes in the Northern Hemisphere, allowing build up of continental ice sheets. Milankovitch recognised that glacials are most severe when the three orbital cycles come together to minimise the amount of solar energy reaching the Northern Hemisphere during summer (resulting in cooler summers). This is because lower temperatures in summer mean that less of the snow that fell during the winter will be melted, allowing snow and ice to build up over many years into large ice sheets.

On the other hand, interglacials occur when the three orbital factors combine to maximise solar energy in the Northern Hemisphere summer (warmer summers). Orbital factors cause increased solar energy input to the Northern Hemisphere during summer when the Earth has a greater tilt than average, and the Northern Hemisphere summer occurs when the planet is nearest the Sun on its orbit. This effect is further enhanced when the orbit is more elliptical than circular. With higher summer temperatures, less of the snow and ice that formed in winter survives through the summer, and over many years this stops the ice sheets from growing, eventually causing them to retreat. The evidence from ocean sediment cores shows that the 96 000-year cycle of eccentricity has been the dominant cycle governing oscillations between glacials and interglacials over the past 900 000 years or so.

## c) Abrupt climatic changes during the Pleistocene

If the Milankovitch theory alone explained climatic changes during the Pleistocene, we would expect long and gradual shifts between glacial and interglacial conditions, just as the changes in solar energy input caused by orbital cycles are long and gradual. However, in the 1980s and 1990s it became clear from studies of past climate that the shifts between glacials and interglacials were remarkably rapid, and that glacials themselves were characterised by strongly fluctuating temperatures rather than being uniformly cold. Evidence contained in ancient layers of ice drilled from the Greenland ice sheet illustrates this best.

Two major ice-drilling projects in Greenland, the European Greenland Ice-core Project (GRIP) and the North American Greenland Ice Sheet Project 2 (GISP2), extracted long cores of ice from the summit of the Greenland ice sheet to depths of over 3000 m. The deepest layers of ice in the cores exceed 100 000 years in age, and annual layers of ice can be identified along the cores extending from the base all the way up to ice at the top that was laid down in recent years. This has made it possible to develop an incredibly detailed picture of how the climate has been changing at Greenland (and in the

North Atlantic region generally) from the present to as far back as the beginning of the last glacial.

Information about past climate has been gained from ice cores in a number of ways.

- The thickness of an ice layer shows how much snow accumulated in a single year.
- Air bubbles trapped within the ice are used to infer the atmospheric concentrations of trace gases, like carbon dioxide and methane, when the ice was formed.
- The acidity of the ice is also measured to identify past volcanic eruptions that caused temporary increases in the amount of sulphuric acid in the atmosphere.
- Past air temperatures are reconstructed by studying the oxygen isotope composition of the ice itself.

As already described, $H_2{}^{16}O$ is more readily evaporated from the sea than $H_2{}^{18}O$. This means that snow falling on the Greenland ice sheet will always have a lower (more negative) $^{18}O/^{16}O$ ratio than sea water. Additionally, colder air is less effective at evaporating $H_2{}^{18}O$ from the sea than warmer air. As the air temperature in the North Atlantic region decreases, the $^{18}O/^{16}O$ ratio of snow falling on Greenland also decreases, becoming even lower (more negative) in relation to the ratio of sea water. Conversely, an increased $^{18}O/^{16}O$ ratio of snow (less negative in relation to sea water) relates to a time of increased temperature. The snow eventually forms a layer of ice that becomes part of the ice sheet, in this way preserving the $^{18}O/^{16}O$ ratio of the snow at the time that it fell.

Changes in the oxygen isotope ratio of layers of ice spanning the GRIP and GISP2 ice cores show that temperatures in the North Atlantic region have been highly variable over the past 110 000 years (Figure 4). Throughout the most recent glacial phase (the **Devensian** in the British Isles), which was well underway by about 73 000 years ago, there have been many large and abrupt climatic changes involving mean annual temperature swings of as much as 8°C occurring within just a few decades. The coldest phases of the Devensian, when $^{18}O/^{16}O$ ratios reach their most negative values on Figure 4, represent average air temperatures in the North Atlantic region that were about 12–13°C lower than the present day. These relatively short phases of intense cold that occurred during the last glacial, and during previous glacials, are known as **stadial** periods. Most of the peaks in $^{18}O/^{16}O$ ratios before the beginning of the Holocene Interglacial 11 500 years ago represent air temperatures that were 5–6°C lower than present, and these brief phases of relative warmth during glacials, lasting 500–2000 years, are known as **interstadial** periods. At about 11 500 years ago the ice core data show a rapid warming to maximum temperatures followed by sustained warmth, with less temperature variation during our current Holocene interglacial.

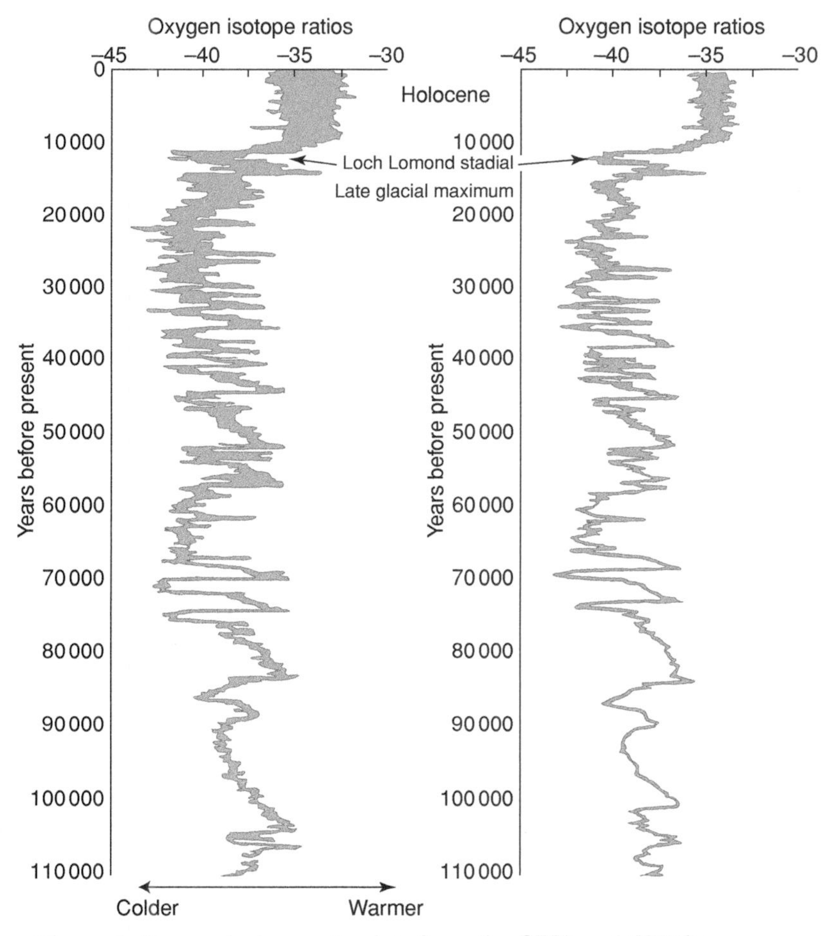


**Figure 4** Oxygen isotope ratio data from the GRIP and GISP2 ice cores plotted against time (more negative values indicate lower temperatures). *Source*: Anderson (2000). Data from the National Snow and Ice Data Center.

During a glacial phase, the stadials represent times when ice sheets increase their size and tundra-type vegetation replaces woodland across mid-latitude land areas in the Northern Hemisphere, whereas the interstadials represent times of temporary ice sheet retreat and the northward spread of woodland. Although the interstadials are phases of relative warmth, they are not of long enough duration to be classed as interglacials. From the ice core data in Figure 4, it is seen that the shifts from cold stadials to warmer interstadials occur more rapidly than the shifts from interstadials to stadials. These abrupt and severe fluctuations in temperature during the last glacial are referred to as **Dansgaard–Oeschger events**, in honour of two ice core scientists who described them.

It is likely that all glacial phases during the Pleistocene have experienced similar climatic fluctuations to those identified from the ice core records for the most recent glacial. These rapid transitions between glacial/interglacial and stadial/interstadial affected the whole planet, although the magnitude of the temperature swings is greatest in the North Atlantic region, including north-west Europe and eastern North America. The cause of these Dansgaard–Oeschger events is thought to involve successive phases of build up and collapse of the large northern ice sheets, as discussed in Chapter 2.

# 5 Late Pleistocene Glaciation

## a) Past glacier ice cover

The maximum extent of glaciers during the Late Pleistocene was reached during the last glaciation between 20 000 and 18 000 years ago when glacier ice covered over 30% of the Earth's land surface. The Antarctic and Greenland ice sheets only covered a slightly larger area than they do now, but there was far more ice across North America and Eurasia. Table 2 shows estimates of Late Pleistocene maximum ice cover and volume for different parts of the world.

**Table 2** Estimates of the area and volume of regions of glacier ice during the Late Pleistocene maximum. *Source*: Smithson *et al.* (2002)

| Region | Est. area ($10^6$ km$^2$) | Est. volume ($10^6$ km$^3$) |
|---|---|---|
| Antarctica | 14.50 | 37.7 |
| Greenland | 2.35 | 8.4 |
| Laurentide ice sheet | 13.40 | 34.8 |
| Cordilleran ice sheet | 2.60 | 1.9 |
| Andes | 0.88 | – |
| European Alps | 0.04 | – |
| Scandinavian ice sheet | 6.60 | 14.2 |
| Asia | 3.90 | – |
| Africa | 0.0003 | – |
| Australasia | 0.07 | – |
| British ice sheet | 0.34 | 0.8 |
| **Total** | **44.68** | **97.8** |

The two major ice sheets of North America, as shown previously on Figure 2, were the **Laurentide ice sheet** to the east and the **Cordilleran ice sheet** to the west. The Laurentide ice sheet extended past the Great Lakes reaching as far south as latitude 39°N during the

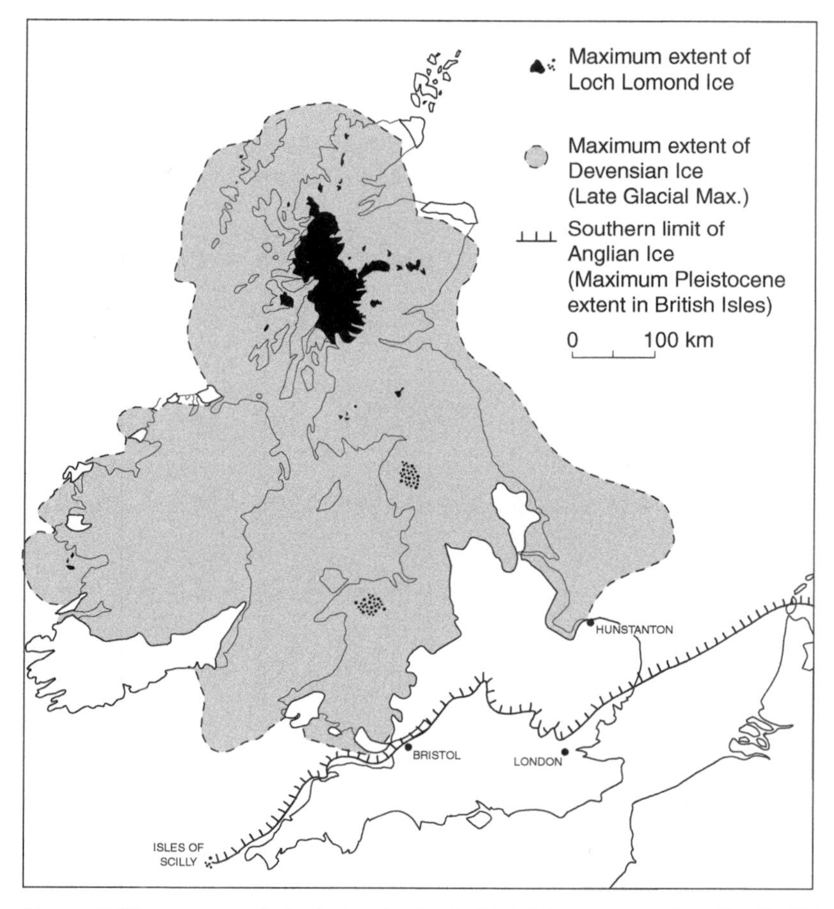


**Figure 5** The extent of glaciation in the British Isles. *Source*: Goudie (1992).

last glaciation. At its centre around Hudson Bay the ice sheet grew to a thickness of nearly 4000 m. The **Scandinavian ice sheet** was thickest over Norway and Sweden, probably exceeding 3000 m in places. It extended eastwards across Russia, merging with glaciers from the Ural Mountains, and to the west it covered much of what is now the North Sea, at times coalescing with the British ice sheet. In the British Isles ice covered most of Ireland, Wales and Scotland and extended over much of northern England and the midlands as far south as East Anglia (Figure 5). There were also large accumulations of ice in Eastern Siberia, in Central Asia, the Andes, New Zealand and in the European Alps.

## b) Effects of the ice cover

With so much water locked up in ice sheets during the last glaciation (over 5% of the world's water) sea level was more than 100 m lower than at present, and large areas of continental shelf that are currently submerged were dry land. For example, there was no English Channel; and Asia and Alaska were connected by the Bering land bridge, which is now the Bering Strait. As the ice sheets retreated, prehistoric people were able to migrate from Asia into North America by crossing this land bridge.

The great weight of ice sheets depressed underlying continental crust and, since glacial retreat, areas of crust have been rising in response to the removal of this overburden. This process of **isostatic uplift** is most pronounced where the ice was thickest. For example, in the eastern Hudson Bay region of Canada, where the Laurentide ice sheet was centred, land is still uplifting at a rate of about 1 cm per year.

The average global temperature was about 5°C lower than today during the height of the last glacial, and average temperatures were more than 10°C lower at high latitudes. Weather patterns across the globe were also very different. For instance, there was abundant rainfall in some regions that are now desert. The expansion of ice sheets in the Northern Hemisphere caused jet streams and mid-latitude depressions to track further south, for example increasing precipitation in the presently arid American South-west. Many other continental regions were drier than they are today.

The northernmost extent of woodland in the Northern Hemisphere was pushed far south during the last glacial maximum. In Europe woodland was restricted to Spain, Italy and the Balkans, and to the north the terrain was open, consisting of steppe and tundra-type vegetation. This landscape provided ideal habitat for large, cold-adapted grazing animals such as the mammoths that became extinct as the last glacial ended. Permafrost extended as far as southern France to the west and the Black Sea to the east, and periglacial processes affected much of Europe (as explained in Chapter 6). In the tropics the climate was cooler and drier, and tropical rainforest was less extensive.

## c) The Lateglacial/Interglacial transition

The ice sheets began retreating about 18 000 years ago, with deglaciation proceeding rapidly by 15 000 years ago as the climate warmed. The period between 15 000 and 12 800 years ago is known as the **Lateglacial Interstadial**. At the beginning of this warm phase, average temperatures in north-west Europe were almost as high as they are today, and woodland began to spread northwards as the ice retreated. However, 12 800 years ago temperature plunged back to glacial conditions, remaining low until about 11 500 years ago when temperature again rose rapidly and the Holocene began. This cold phase at the end of the last glacial lasted approximately 1300 years and is known as the **Younger Dryas event**. During the Younger Dryas the retreat of ice was halted, and glaciers re-advanced in many parts of the world. Glacier ice had almost disappeared from the British Isles during the Lateglacial Interstadial, but during the Younger Dryas an ice field was re-established in the Scottish Highlands. In Britain this phase of glaciation is referred to as the **Loch Lomond Stadial**, and it saw the expansion of cirque and valley glaciers in upland regions of Scotland, England and Wales (Figure 5).

The transition from the Younger Dryas to the Holocene interglacial was characterised by rapid warming and glacier retreat. Data from the Greenland ice cores suggest a mean annual temperature rise of about 7°C in less than 50 years, with most of this warming occurring in as little as 3 years (Figure 4). The pollen from various plants preserved in lake and peat sediments shows that steppe-tundra vegetation was replaced by birch, pine and hazel woodlands across much of north-west Europe within 500 years, followed later by the arrival of temperate deciduous trees such as elm and oak. Sea level also rose as the ice sheets melted, finally stabilising around 7000 years ago once the Laurentide ice sheet had completely disappeared.

## Summary Diagram

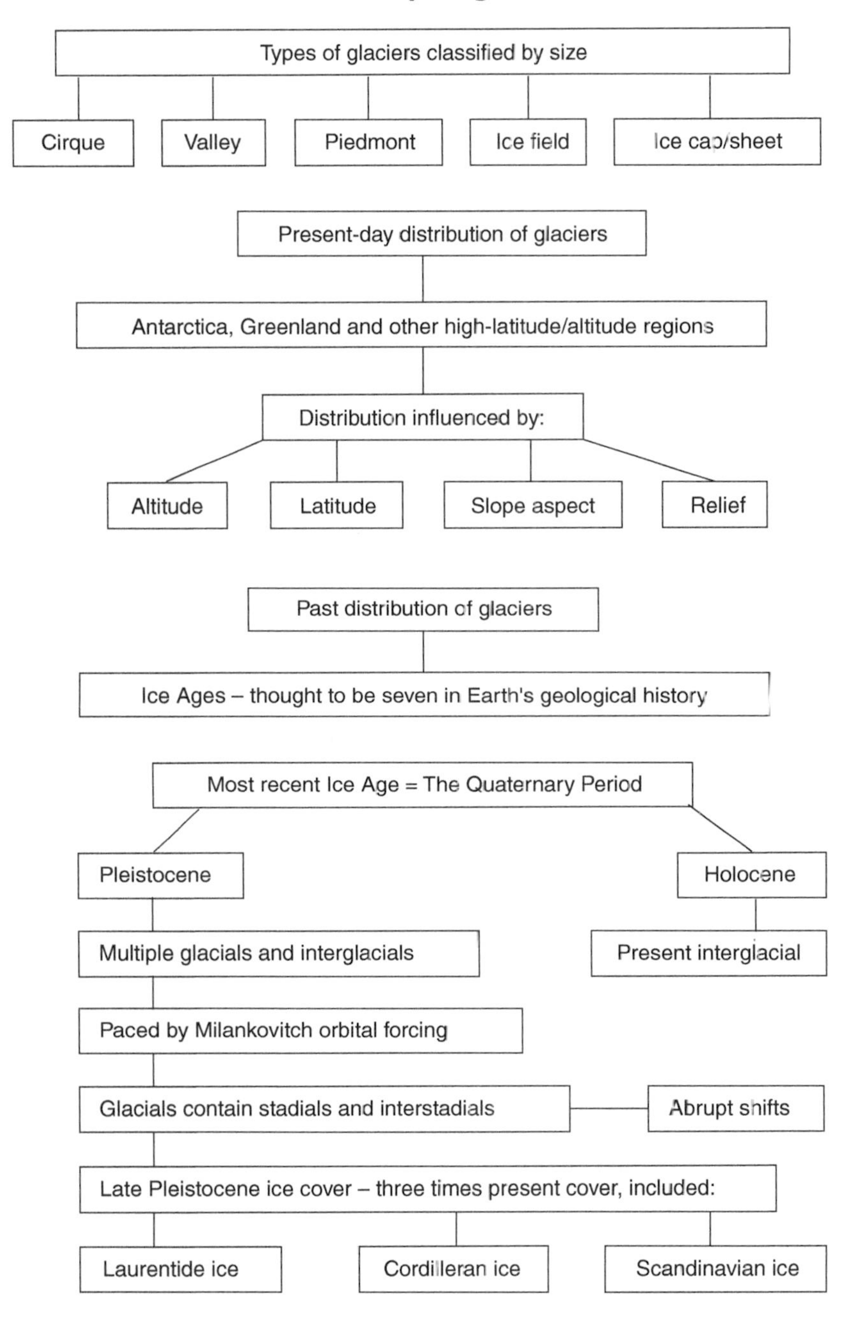

## Questions

1. **a)** Describe how glaciers can be classified by size.
   **b)** To what extent does latitude and altitude affect the distribution of present-day glaciers?
2. Referring to Table 2, determine the percentage contribution of the following ice masses to the total global ice cover during the Late Pleistocene maximum: the *Antarctic ice sheet*, the *Laurentide ice sheet*, the *Scandinavian ice sheet*, the *Cordilleran ice sheet*, the *Greenland ice sheet* and the *British ice sheet*.
3. **a)** Why have there been multiple glacials and interglacials during the Pleistocene?
   **b)** Describe the global effects of glacial/interglacial cycles.
4. **a)** Describe how evidence in ocean sediment cores has enabled reconstruction of glacial/interglacial cycles.
   **b)** Describe how evidence in ice cores has enabled reconstruction of climatic changes during the last glacial.

# 2 Glaciers as Systems

**KEY WORDS**

**Ablation**: loss of mass from a glacier, primarily through melting.
**Ablation zone**: the area of a glacier with a negative net balance (melting exceeds accumulation).
**Accumulation zone**: the area of a glacier with a positive net balance (accumulation exceeds melting).
**Basal thermal regime**: the thermal conditions at the base of a glacier. Glacier ice is 'cold-based' if frozen to the bed and 'warm-based' if meltwater is present.
**Glacier mass balance**: the balance between glacier accumulation and ablation.
**Net balance**: the glacier mass balance averaged over a year.
**Pressure melting point**: the slightly lower temperature at which ice melts caused by increased pressure (weight) of overlying glacier ice.
**Regelation**: the refreezing of meltwater produced by pressure melting beneath a glacier. This is an important process in glacial movement.

## 1 The Formation of Glacier Ice

The ice that makes up glaciers and ice sheets is derived primarily from compacted snow, with some contribution from other forms of precipitation such as sleet, hail and rain, which freezes on top of and inside the glacier. The first stage in the formation of a glacier is the accumulation of a permanent snowfield that survives through the summer and results in net growth averaged over the whole year. This will occur at high altitudes in the mountains or in high-latitude regions where summer temperatures are low so that winter accumulation of snow exceeds summer snow melt. A key climate factor for the formation of glacier ice in a particular place is the number of **positive degree-days**. This refers to the number of days in the summer when temperatures are greater than 0°C. Fewer positive degree-days means less snow melt, increasing the likelihood that any snow that fell during the winter will survive the summer to cause net accumulation of snow over the year.

After successive years of snowfield growth, lower layers of snow become increasingly compressed due to the weight of overlying snow. Dry fresh snow has a density of about 0.1 g $cm^{-3}$. Once snow becomes compacted to a density of around 0.5 g $cm^{-3}$, it exhibits a granular, interlocking structure and is known as **firn** (also termed **névé**). The increase in density is aided by pressure-induced melting of snow at depth followed by refreezing of water to fill the gaps between individual ice crystals. As the snowfield grows, this process continues until the deeper layers of firn are transformed into glacier ice with a density of about 0.9 g $cm^{-3}$. With climatic conditions of high snowfall and the temperatures oscillating above and below freezing, the effect of surface snow melt with subsequent downward percolation and refreezing of water within the gaps between crystals can cause glacier ice to form at depths of little more than 10 m beneath the snow surface. Under these conditions, as exemplified in glaciers of Alaska's Coast Range, glacier ice can form from snow within 5 years of burial. In contrast, within the much colder and drier interiors of Antarctica and Greenland, the transformation from snow to glacier ice is not complete until the snow is buried to depths of 60 m or more – taking hundreds, and in some places even thousands, of years.

Under the heavy weight of the overlying snowfield, the glacier ice formed beneath begins to deform and move outwards away from the centre of greatest pressure. On a mountainside this will, of course, be downhill; but in a relatively flat area the motion can be outwards in all directions in the form of a spreading dome, analogous on a large scale to the motion of ice away from the centre of the Greenland ice sheet. Once glacier ice is moving away from the centre of snow accumulation, what began as a snowfield has been transformed into a glacier.

## 2 The Glacier as a System

Regardless of their size, all glaciers can be viewed as systems, and all glaciers work in essentially the same way. Systems are characterised by having inputs and outputs, as well as processes that store, transport and transform the matter and energy going through the system. As systems, glaciers are also dynamic – constantly adjusting to climate-induced changes in the balance between inputs and outputs.

### a) Inputs and outputs

The material input into the system is mostly snow, either through direct snowfall or snow avalanching down onto the glacier from surrounding slopes. The increase in the mass of ice caused by the addition of snow is referred to as glacier **accumulation**. Mass is lost from a glacier mainly by surface melting, but also by melting of ice inside and at the base of the glacier resulting in an output of meltwater. There can also be loss of mass by evaporation (sublimation of ice directly to water vapour), by the action of wind blowing snow off the glacier and by **calving**, which involves icebergs breaking off a glacier where it borders the sea, a lake or a stream. The loss of mass from a glacier is referred to as **ablation**.

Glaciers are also characterised by the input and output of rock debris. The input of rock debris comes from weathering and erosion of slopes above the glacier and from the erosive action of the glacier itself. Rock debris, as well as any soil and organic material incorporated into the glacier system, is then transported and eventually deposited, becoming an output of the glacier.

As with any other system, glaciers also have inputs and outputs of energy. The Earth's climate is powered by the Sun, and glaciers would not exist if it were not for solar energy evaporating water from the oceans to create the moist air masses that produce snowfall at high altitudes and latitudes. Solar energy, and in some cases geothermal energy, are important for melting glacier ice, the former causing surface melt and the latter contributing to melt at the base of a glacier. Furthermore, the energy used in the phase change from ice to liquid water becomes stored in the meltwater, and can be released as latent heat within a glacier if the meltwater refreezes. In addition to solar and geothermal energy, there is an input of potential energy due to gravity, and this energy is transformed into kinetic energy through the movement of glacier ice downhill.

## b) Glacier mass balance

The inputs to and outputs from a glacier are not constant, but vary continually over both short and long timescales. The glacier system constantly adjusts to changes in the balance between accumulation and ablation, and this is reflected in the **mass balance** of a glacier. If accumulation exceeds ablation, as is normally the case in the winter, then a glacier gains mass therefore having a **positive mass balance**. In the summer when there is more ablation than accumulation, a glacier has a **negative mass balance** and loses mass. Over the course of a year, therefore, glaciers expand and contract as the mass balance changes with the seasons.

Whether a glacier grows or shrinks over a longer period depends upon the mass balance averaged over a year, known as the **net balance**. The way in which the mass balance changes seasonally is illustrated in Figure 6. If accumulation in the winter is equalled by ablation in the summer, then the annual net balance is zero and the glacier has neither increased nor decreased in size for the year as a whole. The **balance year** for calculating the net balance is taken from the time of minimum mass in one year to the time of minimum mass in the next year. This is from autumn to autumn because the glacier mass will have reached its minimum for the year following the period of summer ablation. For a glacier to grow over the long term, the net balance for most years must be positive. In other words, in most years accumulation in winter must be greater than ablation in summer. A glacier retreats, and may eventually disappear, if its annual net balance is negative over many years.

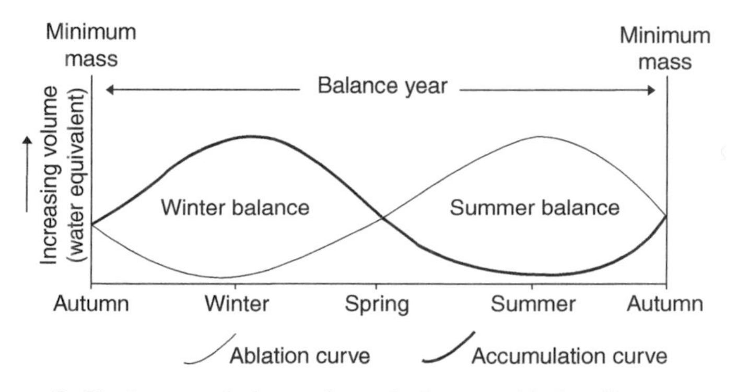


**Figure 6** Glacier mass balance through the year (darker line represents accumulation; lighter line, ablation). *Source*: Sugden and John (1976).

## c) The accumulation and ablation zones

The discussion so far has focused on the net balance of a glacier in its entirety without considering the differences in the balance between accumulation and ablation that exist across a glacier. Using an alpine valley glacier as an example, the origin of the glacier high in the mountains is the area with the most positive net balance over the year because snowfall is high, and a low mean annual temperature means that there is little melting. This high elevation zone of the glacier is known as the **accumulation zone**. Toward the **snout** (or end) of the glacier at lower elevations the average annual temperature is higher, causing more ablation. There is also less precipitation in the form of snow over the course of a year, causing less accumulation than higher up. Therefore, this area of the glacier, the **ablation zone**, has a negative net balance. Glacier ice is continually produced in the accumulation zone, and it is always moving toward the ablation zone where it becomes an output from the glacier either through melting, sublimation or calving.

The net balance is at its most positive at the top of the glacier and most negative at the snout, whereas at middle elevations accumulation and ablation over the course of a year are nearly equal. The point along a glacier where annual accumulation and annual ablation balance each other exactly (a net balance of zero) is called the **equilibrium line**, or **firn line**. In the case of an ice cap or ice sheet, the net balance is most positive in the central area of the mass of glacier ice and most negative along the periphery, bordering on either land or sea (Figure 7).

If the positive net balance in the accumulation zone is equal to the negative net balance in the ablation zone, then the net balance for the glacier as a whole is zero. Such a glacier is in a **steady state** because it

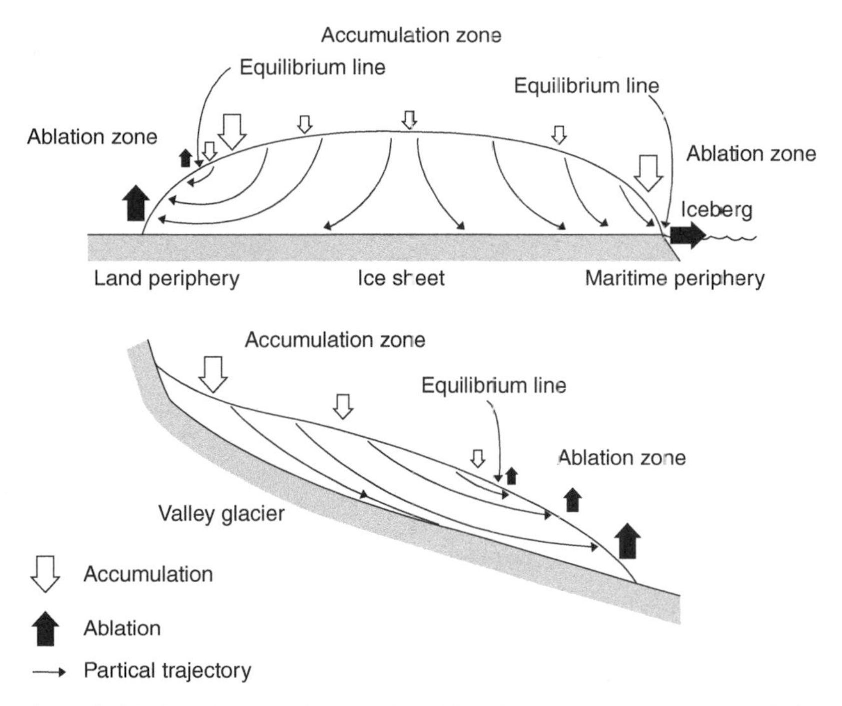


**Figure 7** Model of an ice sheet and a valley glacier illustrating accumulation and ablation zones. *Source*: Bennett and Glasser (1996).

has neither gained nor lost mass over the balance year. However, this does not mean that the margin of the glacier remains stationary through the year. In the winter the glacier gains mass because there is more snowfall at lower elevations and less melting near the ice margin. The accumulation zone is enlarged relative to the ablation zone, and the snout advances. Conversely, in the summer the ablation zone expands, the glacier loses mass and the snout retreats. However, under steady state conditions, the snout will finish the balance year in autumn in the same position as in the previous autumn.

If a change in climate results in a balance year that is not zero, the glacier will either retreat or advance relative to the previous year. Taking the example of a positive net balance for a valley glacier, this will cause the accumulation zone of the glacier to expand and the equilibrium line and glacier snout to move downward to a lower elevation. Studies of glaciers in many parts of the world have revealed periods of expansion and retreat in recent history as climate changes have shifted the net balance to either positive or negative over a timescale of decades to centuries.

## CASE STUDY: HISTORICAL CHANGES IN GLACIER MASS BALANCE

Although the present interglacial, the Holocene, has been a time of comparative warmth relative to the last glacial, the climate has not remained constant. Instead, climatic changes during the Holocene have influenced existing glaciers throughout the world, sometimes causing them to expand and other times to retreat. In historic times, one of the most important phases of glacier advance began in the 14th century AD when mean annual temperatures in most parts of the world declined from a peak during the Middle Ages (the Medieval Warm Period). The period spanning the 14th to the 19th centuries was punctuated by a number of cold spells that, at their most severe, represented times when the mean annual temperature in Europe was between 1 and 2°C lower than the 20th century mean. Cooler temperatures and decreased ablation in summer caused alpine glaciers in most mountainous parts of the world to extend further down their valleys, and the margins of the Greenland and Antarctic ice sheets were extended. This phase of glacier expansion has been termed the **Little Ice Age** because it was significant worldwide, but at a far smaller scale than expansion during a glacial.

Mean annual temperatures reached their lowest values during the Little Ice Age between 1550 and 1750 AD, and again in the mid-19th century. In Norway, where it has been possible to reconstruct historical changes in the size of glaciers in great detail, it has been found that this phase of glacier expansion reached its maximum around 1750 AD. In Iceland, expansion of glaciers and ice caps also culminated about 1750, and, following a period of stagnation or retreat, there was another advance culminating around 1850. The pattern is similar, although not exactly the same, in the European Alps and in North America. For example, the Rhône Glacier in Switzerland had strong phases of expansion between 1600 and 1680, 1770 and 1780, and between 1835 and 1855. In Alaska many glaciers reached maxima between 1700 and 1835.

In contrast with the cooler conditions of the 19th century, over the 20th century it is estimated that the Earth's average surface temperature has risen by about 0.6°C. While this temperature increase is probably partly due to natural factors, climate scientists believe that increasing concentrations of carbon dioxide and other greenhouse gases in the atmosphere caused by humans is also responsible. This recent warming has greatly increased summer ablation, causing around 75% of all glaciers across the world to have a negative annual net balance. While the

Little Ice Age was a period of predominantly positive net balance causing glacier expansion, through the 20th century and now into the 21st, glacier retreat has been accelerating. Remarkably, the Alps have lost approximately half the glacier ice that was present in 1850 AD, and, if current trends continue, by 2050 ice cover in the Alps may be only a fifth of what it was before the industrial revolution. Evidence for rapid glacier retreat is abundant throughout the Alps. For instance, many mountaineering huts in the Alps that were built to be easily accessible from the glaciers early in the 20th century are now perched high above them on steep slopes.

The situation is similar in many other parts of the world. In the eastern Himalayas of Asia some 2000 glaciers have disappeared in the last 100 years, and the large Pindari Glacier in the Himalayas is retreating at a rate of about 135 m per year. In the Rocky Mountains of the USA, the number of existing glaciers has declined from 150 to fewer than 50 since 1850 AD, and Alaska's Columbia Glacier has retreated by over 13 km since 1982. In South America the Quelccaya Ice Cap in the Andes of Peru has been retreating at a rate of around 30 m per year through the 1990s. Also in the Southern Hemisphere, glaciers on Mount Kilimanjaro in Tanzania have lost around 70% of their mass since the late 1800s, and glacier retreat on Mount Kenya has been even more severe. There are no clear trends as yet for the Greenland and Antarctic ice sheets, although the former will probably lose mass over the next century while the latter may gain mass because of increased precipitation.

## 3 Cold- and Warm-based Glaciers

Glaciers vary widely in temperature. Some glaciers have an average temperature at or close to 0°C, while in the polar ice sheets the temperature can be far lower. Glacier ice in parts of the Antarctic ice sheet is as cold as –40°C. In general, the temperature of glacier ice is close to the mean annual temperature of the air in the area where it formed. This is caused by the compaction and mixing within the glacier of many individual layers of snow and trapped air of differing temperatures representing snowfalls that occurred throughout a year. As glacier ice increases in thickness and moves, the temperature characteristics of the glacier can change. At a thickness of more than 500 m the temperature of ice increases with depth because of the insulation of geothermal heat at the base of the glacier. The temperature of ice within a glacier can also be increased through frictional heat released as glacier ice moves, and by the release of latent heat when meltwater within a glacier refreezes.

Glaciers can be classified as either **cold-based** or **warm-based** depending upon whether or not they are frozen to the underlying bedrock. Cold polar glaciers, particularly in Antarctica and Greenland, tend to be cold-based because the average temperature of the ice is well below 0°C, and the accumulation of geothermal heat is usually not great enough to raise the temperature at the base of the glacier to the melting point. There is also relatively little surface melt during the cool and short polar summer. As a result, little meltwater is available to percolate downward through the ice and contribute latent heat on refreezing. Very low temperatures from the surface down to the base cause the glacier to be permanently frozen to its bed, and meltwater beneath the glacier is absent.

Outside of the polar region, most glaciers are of the warm-based, temperate type. The temperature in the surface layer of this type of glacier fluctuates above and below melting point depending on the time of year, whereas the temperature of the rest of the ice extending downward to the base is close to the melting point. The melting point of ice is 0°C near the surface, but with increased pressure of overlying ice at deeper levels within a glacier, water can exist as a liquid at temperatures below 0°C. For example, the **pressure melting point** beneath 2000 m of ice is –1.6°C. As glaciers grow and thicken, the lowering of the pressure melting point beneath the glacier causes **basal ice** (the lowest layers of ice within the glacier) to melt continually, and this prevents the glacier from being frozen to its bed. In addition to the effect of pressure, the contribution of geothermal energy and the percolation of meltwater downward through the ice also prevents the glacier from freezing to its bed.

Glaciers cannot always be neatly divided into the warm-based or cold-based categories. For example, large glaciers and ice caps can be cold-based in their upper regions and warm-based near their margins when they extend across significantly different climatic zones. Differences in ice thickness and the shape of the underlying bed across a glacier can also cause differences in the amount of insulation and pressure near the base, causing some areas of basal ice to melt while other areas remain frozen to the bed. The temperature conditions beneath a glacier (the **basal thermal regime**) can show much spatial variation.

## 4 Glacier Ice Movement

The fundamental cause of glacier ice movement is the imbalance between accumulation and ablation across a glacier. As mass builds up over time in the accumulation zone, the weight of the snow and ice exerts an increasing down-slope force due to gravity termed shear stress. As the snow and glacier ice accumulates, the slope eventually becomes steep enough that underlying ice can no longer support it (shear stress increases as the slope angle increases). Once the shear stress is great enough to overcome resisting forces of ice strength and friction (the shear strength), the glacier ice deforms and moves downward away from the zone of accumulation. The ice movement toward the ablation zone prevents further build up of the slope angle in the accumulation zone thereby maintaining the glacier at an equilibrium slope angle.

As long as there is an imbalance between accumulation and ablation across a glacier a shear stress will be produced, always causing glacier ice to move from the area of excess accumulation to the area of excess ablation. Glacier ice is therefore always moving forward, toward the margin or snout, regardless of whether the glacier as a whole is advancing or retreating.

The speed at which glacier ice moves forward depends on the degree of imbalance, or the gradient, between the zone of accumulation and the zone of ablation. Glaciers in temperate-maritime climates have greater snowfall in winter and experience more rapid ablation in summer than glaciers of colder and more continental climates. Therefore, the imbalance between the accumulation and ablation zones is greater in the former type, and glacier ice must move more rapidly towards the ablation zone to maintain the equilibrium slope angle. On the other hand, the slow rates of accumulation and ablation characteristic of glaciers in cold, continental climates result in a smaller gradient and slower ice movement.

Glacier ice moves in response to the shear stress acting upon it in two main ways – by **internal deformation** and **basal sliding**. Depending on the nature of the substrate on which the glacier rests, glacier ice can also move through the process of **subglacial bed deformation**.

## a) Internal deformation

This process involves the response of individual grains of ice within the glacier to the overburden pressure. The degree of deformation of the grains of ice (rate of internal deformation) increases with increased ice thickness and/or surface slope angle. Ice also deforms more easily at higher temperatures. The deformation of the ice in response to stress is called **ice creep**, and this involves both the elongation of individual ice crystals and the displacement of ice crystals relative to each other. This movement of ice occurs along slip planes that form within, and between, individual ice grains. Although the response of the ice to the shear stress occurs at a very small scale, displacement of ice crystals throughout the whole mass of the ice translates into large-scale motion. This type of motion can be thought of as 'ice flow', even though the nature of the flow is unlike the flow of a liquid.

In some parts of the glacier mass, ice creep cannot adjust quickly enough to the applied stress and **ice faulting** occurs. This results in the separation and displacement of large sections, or blocks, of ice relative to each other. Where the fractures (faults) between sections of ice appear at the glacier surface they are known as **crevasses**.

The velocity of ice movement by internal deformation and the development of crevasses are strongly influenced by the slope gradient of the glacier, as shown in Figure 8. Where the slope gradient is increased, there is an acceleration of ice and **extensional flow**. Such conditions occur in the accumulation zone and where the underlying bedrock becomes steeper producing an **ice fall**. Where there is extension, normal faults are produced as blocks of ice fracture and slip relative to each other, opening up crevasses of the transverse type. **Transverse crevasses** cut across the glacier at right angles to the direction of glacier flow and can be several metres wide. Near the ablation zone, or where the slope angle of underlying bedrock is reduced, the ice decelerates and there is **compressional flow**. The compression of ice causes the formation of reverse faults as sections of ice from up-glacier are pushed and thrust upwards against lower sections of ice. In these areas crevasses close as ice is compressed.

In addition to the effect of slope gradient, changes in the width of a valley can also cause ice fracturing. For example, longitudinal crevasses that are oriented parallel to the flow direction form as ice spreads out laterally where a valley widens. At the snout of a glacier, radial crevasses can form a splayed pattern of fractures where ice spreads out into a broad lobe. The effect of friction on the sides of a valley slows ice movement relative to the ice near the middle of a glacier, and this can create fractures that form marginal crevasses along the sides of a glacier.

## b) Basal sliding

The movement induced by basal sliding relates to the presence of meltwater beneath a glacier. This type of ice movement applies to warm-based glacier ice, but cannot occur where a glacier is frozen to its bed (cold-based). The meltwater acts as a lubricant reducing friction with the underlying bedrock. The bed friction relates to the number of points of contact between the ice and the bedrock, and even just a few millimetres thickness of water beneath a glacier can significantly reduce the amount of contact, greatly increasing rates of basal sliding. The contribution of basal sliding to total ice movement can vary widely from about 20 to 80%, and it is most effective on relatively steep slopes in summer when there is the most meltwater.

There are two specific processes by which glaciers can slide over their beds, the first being by **enhanced basal creep**. Not to be confused with the ice creep process already discussed, enhanced basal creep explains how basal ice deforms around irregularities on the bedrock surface. The underlying bedrock is never completely smooth. Rather it is usually rough and bumpy, with various rocks and

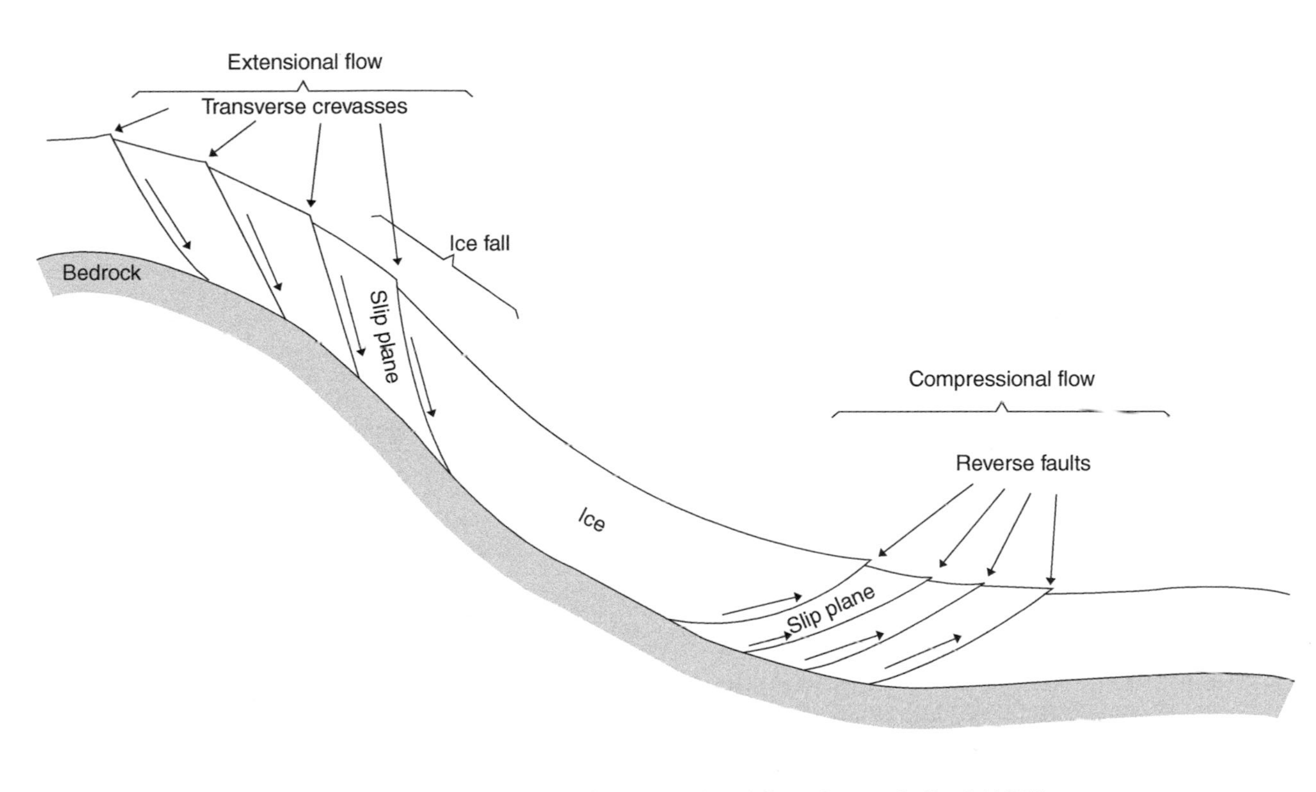


**Figure 8** Extensional and compressional flow. *Source*: Collard (1988).

boulders that protrude into the overlying ice. As a glacier moves over its bed, the pressure on basal ice will increase on the up-glacier (upstream) side of an obstacle causing increased ice deformation. The ice adjusts to this increased pressure by moulding itself around the obstacle; and the larger the obstacle, the greater the rate of deformation.

The second process of basal sliding is by **regelation slip**. If the temperature of the basal ice is near the pressure melting point, then the increased pressure on basal ice on the up-glacier side of a bedrock obstacle may induce melting. The meltwater then flows around the obstacle to the down-glacier side where the pressure is lower and the meltwater refreezes. When glacier ice melts under increased pressure and then refreezes as a result of a reduction in pressure, it is termed **regelation ice**. With this process repeating itself around numerous bedrock obstacles, there is large-scale slipping of the glacier over its bed.

## c) Subglacial bed deformation

When a glacier moves over relatively weak or unconsolidated sediment instead of hard rock, the sediment itself can deform under the weight of the glacier, moving the ice along with it. This occurs when pore water pressure inside the sediment is high, because of both the pressure of overlying ice and the addition of glacial meltwater to the total water content of the sediment. High pore water pressure within the sediment reduces the friction between individual grains, causing the sediment to deform and flow forward in response to the glacier overburden. In Iceland, observations of glaciers and studies of the underlying material (subglacial bed) have shown that this process can account for up to 90% of the forward motion of glacier ice.

## d) Glacier ice velocity

The total forward motion of glacier ice comes from a combination of the processes described above. All glaciers are subject to internal deformation by ice creep and ice faulting. In a valley glacier, the forward velocity of internal deformation is greatest away from the bed and valley sides where friction is greatest (Figure 9a). In the case of a cold-based glacier resting on hard bedrock, internal deformation will be the only component of ice movement. A warm-based glacier resting on hard bedrock is subject to both internal deformation and basal sliding (Figure 9b). The component of forward velocity produced by basal sliding is equal throughout the ice mass because it involves the sliding of large sections of ice along the bed and valley sides. Warm-based glaciers have a greater overall velocity of ice movement than cold-based glaciers because of the addition of basal sliding. Even greater forward velocities are reached when a warm-based glacier

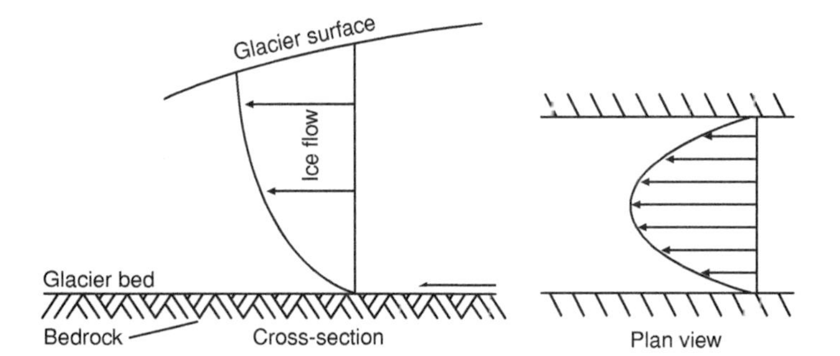


(a) Cold-based glacier resting on bedrock

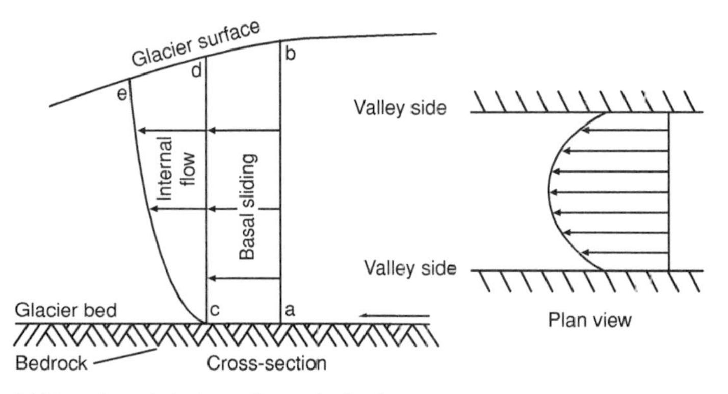


(b) Warm-based glacier resting on bedrock

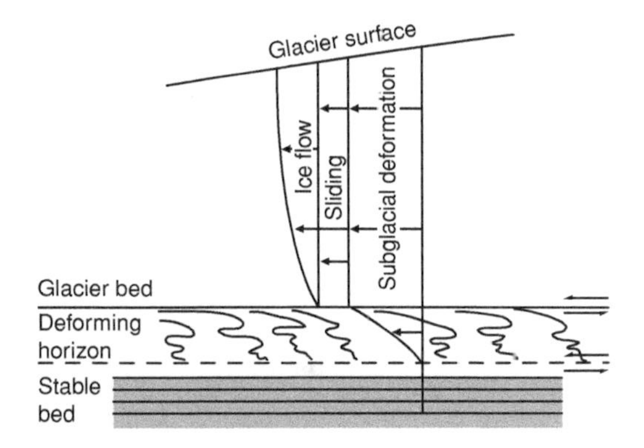


(c) Warm-based glacier resting on deformable sediment

**Figure 9** Components of glacier motion.
*Source*: Duff and Duff (1993).

moves over deformable sediment (Figure 9c). As with basal sliding, the component of forward velocity produced by subglacial bed deformation is also equal throughout the ice mass.

Observations of glaciers across the world have shown a great deal of variation in the total velocity of glacier ice, with most glaciers having velocities between 3 and 300 m per year. In some cases where the relief is steep and there is rapid snow accumulation, velocities can reach 1–2 km per year. Large outlet glaciers, also known as ice streams, along the margins of the Greenland and Antarctic ice sheets can have velocities of up to 12 km per year. These are areas where the movement of vast quantities of ice outwards from the ice sheets is under great pressure and highly concentrated. The highest velocities occur when glaciers **surge**. Surging refers to the periodic collapse of a glacier when the mass and slope angle of ice builds up to a critical level within the accumulation zone. During a surge, ice races forward at velocities between 10 and 100 times the normal velocity. After the surge, glacier ice movement returns to normal velocities and the ice mass gradually thickens up-glacier until the critical mass for a surge is again reached. Glacier surges are relatively rare, and it is estimated that only about 4% of all glaciers today are prone to surging. However, glacier surging on a very large scale appears to have been an important feature of Pleistocene glacials. It has been proposed recently that the rapid climatic changes during the last glacial revealed in the Greenland ice cores (Dansgaard–Oeschger events) were related to large-scale, periodic surging of the Northern Hemisphere ice sheets.

## 5 Ice Sheet Growth and Collapse

Analyses of ocean sediments in the North Atlantic have shown that there were huge discharges of icebergs from the margins of the northern ice sheets around the times when climate switched from the coldest phases of stadials to rapid interstadial warming (see Figure 4). These periods of iceberg discharge into the Atlantic Ocean are known as **Heinrich events** (after their discoverer) and indicate times when the ice sheets became unstable, surged forward and began breaking up along their ocean margins sending huge armadas of ice floating southward.

The coldest temperatures reached during the stadials were probably caused by the Heinrich events because of the way that icebergs and meltwater would have interfered with the transport of warm sea water northward by the Atlantic Ocean's surface currents, the Gulf Stream and North Atlantic Drift. Once the effect of icebergs had run its course, and the ice sheets were significantly reduced in size, the warm ocean currents in the North Atlantic were able to take effect (switching back on), rapidly increasing temperatures in the North

Atlantic region to interstadial conditions. Warmer conditions would then have brought additional moisture to high latitudes causing more snowfall and enabling ice sheets to build back up again, thereby cooling the climate back towards stadial conditions. The cycle is thought to have repeated itself as ice sheets again reached such a large size that they became unstable, leading to another phase of ice sheet collapse around the North Atlantic.

The precise reasons for periodic ice sheet collapse during the Pleistocene are still the subject of much research, but it appears to be the result of a combination of internal and external factors. As ice sheets thicken, there is a greater insulation effect on the temperature of ice at the base. This causes geothermal and frictional heat to build up under the ice sheet, particularly around the margins of the ice sheet where it is moving over weaker rocks and sediments. Once the basal temperature is high enough, and the glacier ice is fully warm-based, sediments beneath the ice sheet thaw. The extra meltwater produced reduces the friction beneath the ice causing huge volumes of ice to begin surging and breaking away from the ice sheet, greatly reducing its size. In addition to these internal processes, there must also be external factors involved because separate ice sheets on both sides of the Atlantic Ocean, as well as small ice sheets in the Andes and New Zealand, appear to have built up and then surged synchronously. The external factors are not understood fully, but probably involved ice sheets around the world influencing, and being influenced by, global ocean circulation and fluctuations in the atmospheric concentration of greenhouse gases.

## Summary Diagram

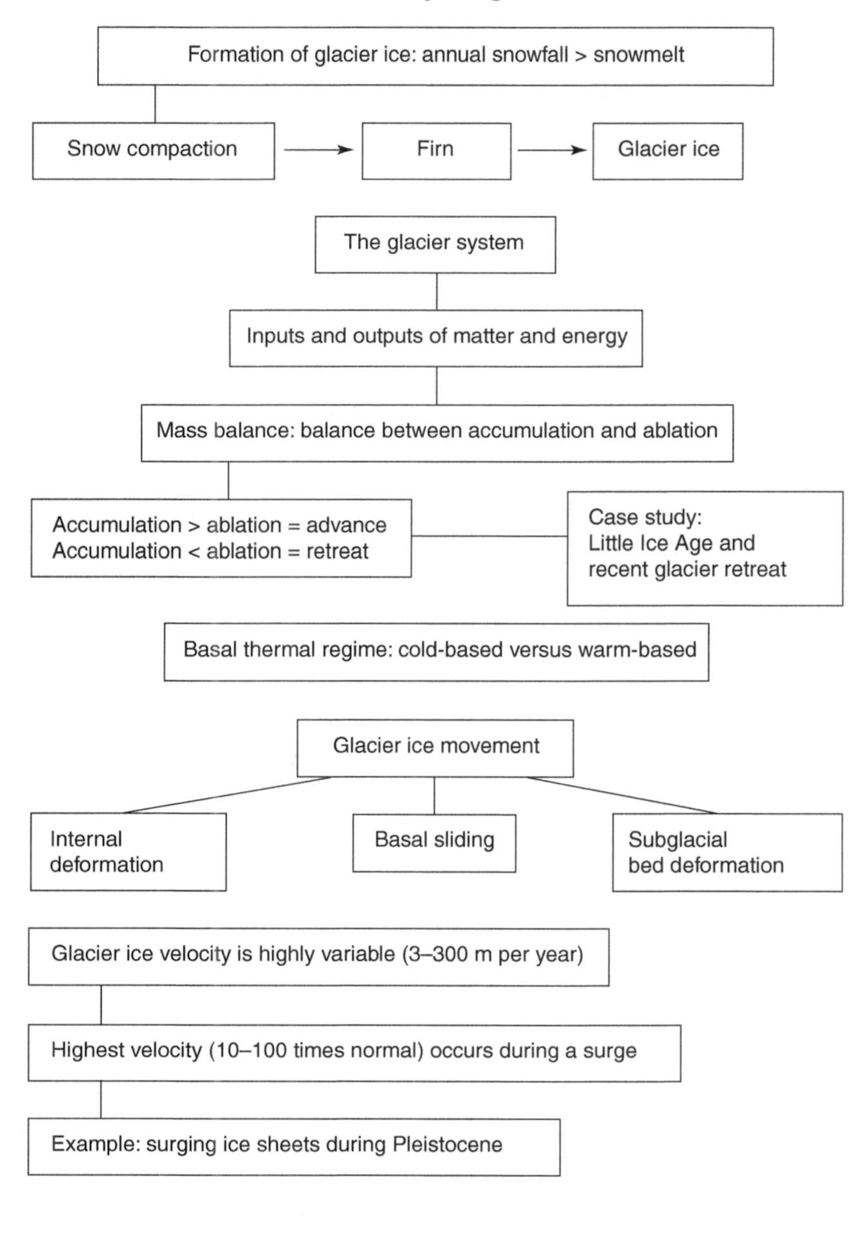

## Questions

1. **a)** Explain how glacier ice forms.
   **b)** Explain what is meant by the accumulation zone, ablation zone and equilibrium line of a glacier.
2. **a)** How does the net balance determine whether a glacier advances, retreats or remains stationary?
   **b)** What was the 'Little Ice Age'?
3. **a)** Describe the difference between 'cold-based' and 'warm-based' glacier ice.
   **b)** Explain why warm-based glacier ice has a higher velocity than cold-based glacier ice.
4. **a)** Summarise the three main mechanisms of glacier movement (internal deformation, basal sliding and subglacial bed deformation).
   **b)** What is meant by a glacier surging?

# 3 Glacial Erosion

**KEY WORDS**

**Entrainment**: the process by which material from the glacier's bed becomes incorporated into the moving ice.
**Glacial abrasion**: the erosional process whereby rock material frozen to the underside of a glacier abrades underlying rock as the glacier moves over its bed.
**Glacial plucking**: the erosional process whereby glaciers loosen, detach and entrain chunks of rock from their beds.
**Nivation**: the formation and enlargement of a hollow on a hillside that may evolve into a cirque through the combined action of freeze–thaw weathering, mass movement and meltwater erosion beneath and around a snow patch.

## 1 Processes of Glacial Erosion

Glaciers modify the landscape over which they move profoundly. Through processes of erosion glaciers detach and carry away rock and sediment, eventually depositing the material elsewhere. The erosion and removal of underlying material by glaciers deepens valleys and sharpens ridges and peaks, leaving behind more dramatic scenery. The precise way in which glaciers erode is not perfectly understood because of the difficulties in observing processes that occur beneath the ice. However, from studying the resultant landforms of erosion it has been possible to achieve some understanding of how these processes work. The processes of erosion can be divided into three main categories – **glacial abrasion**, **glacial plucking** and **glacial meltwater erosion**.

### a) Glacial abrasion

Glacial abrasion involves the effect of rock material at the base of a glacier being dragged across the bedrock surface. Rocks of various sizes frozen to the underside of the ice act like sandpaper, scratching and scouring the underlying rock and also removing less resistant material. In addition to deepening valleys and wearing down the land, glacial abrasion also leaves behind a number of small-scale features on rock surfaces, such as **striations** and **chatter marks**. As bits of bedrock are chipped off, additional rock material may become frozen onto the base of the moving ice (**entrained**), thereby increasing the amount of abrasion. Furthermore, abrasion grinds down rock material producing very fine-grained **rock flour** (grain sizes under 0.1 mm in diameter) that is transported away from the glacier in meltwater. It is the high concentration of rock flour held in suspension within meltwater that gives the characteristic blue-green colour to streams and lakes found along the margins of glaciers. Studies of glacial abrasion in Iceland and in the French Alps suggest that average rates of abrasion range from as little as 1 mm per year to over 30 mm per year.

## b) Glacial plucking

As distinct from the scratching and chipping away of bedrock caused by abrasion, glacial plucking, also referred to as **glacial quarrying**, is the mechanism by which glaciers detach and remove large chunks of rock from their beds. It is a two step process, involving first the fracturing of bedrock beneath the glacier followed by **entrainment** of the loosened rock by the ice. The process of entrainment refers to loose rock material being frozen onto the base of the glacier, thereby being picked up and incorporated into the glacier ice.

Glacial plucking exploits pre-existing fractures within rock that may also have been deepened and widened by freeze–thaw weathering before the arrival of the glacier. As glacial erosion proceeds, the removal of overlying material also causes pressure release (dilatation), producing fractures in rock parallel to the erosion surface as the bedrock adjusts and expands due to unloading. Replacement of rock by ice reduces the total weight on underlying rock because ice is approximately one-third the density of rock. However, the glacier itself applies pressure to the bedrock, and the amount of pressure changes with time and varies along the course of the glacier. This sets up different stresses in different areas of the bedrock causing rock to fracture along lines of weakness. The presence of meltwater under pressure beneath the glacier also plays a key role in widening joints and creating new fractures. Water reduces friction along slip planes within rock, and variations in water pressure beneath the ice will alter the degree of downward pressure (normal stress) of the glacier on the bedrock surface, again helping to vary the stresses and propagate fractures along lines of weakness. Together, all of these factors contribute to the weakening and loosening of bedrock beneath the glacier.

The second step of glacial plucking, the entrainment of rock, only occurs once the creation and widening of fractures has provided a supply of loose material. Like the fracturing of rock, entrainment also results from a number of interrelated factors. Most importantly, variations in pressure beneath the ice create localised areas of pressure melting and other areas of regelation. The concept of changes in the temperature at which ice melts caused by changes in pressure (the pressure melting point) is explained in Chapter 2. On the down-glacier side (lee side) of a rock obstacle pressure is less, and this is where a piece of loosened rock can be pulled out of the bed as regelation of surrounding meltwater causes the rock to become frozen to the base of the ice. The detaching or 'plucking' of rock is also aided by the frictional drag of the ice itself as it moves over the bed. Moreover, the presence of meltwater toward the up-glacier side of a rock obstacle, where ice pressure is greater, reduces the frictional resistance that holds the rock in place along the fractures and helps to push loose rock away from the bed.

### c) Subglacial meltwater erosion

Meltwater beneath the glacier causes erosion through both mechanical and chemical means. This form of erosion increases in importance toward the snout of a glacier where there is the most meltwater. Fluvial abrasion occurs where meltwater flows across the bedrock surface under pressure. This is the process by which particles carried in the flow abrade the bedrock, causing scouring and grooving. This is particularly effective if the meltwater is carrying a relatively coarse load of suspended sediment with a high proportion of particles sized between 0.63 and 2 mm diameter. In areas where meltwater velocity is very high and the channel is rough and irregular, fluvial cavitation also contributes to the mechanical erosion. Where meltwater flow accelerates around obstacles on the bed, areas of low pressure can form causing water to vaporise and form vapour bubbles. These bubbles of vapour eventually collapse as pressures increases, causing shock waves that can be very effective in opening up microscopic cracks in the rock and loosening mineral grains.

Depending on the type of bedrock, glacial meltwater can also cause erosion through dissolving minerals and carrying away the solutes. This is particularly effective in areas of limestone or chalk because of their susceptibility to carbonation weathering. Carbon dioxide is more soluble in water at lower temperatures, and this causes glacial meltwater to contain a relatively high concentration of carbonic acid. In carbonation weathering, carbonic acid reacts with calcium carbonate (the dominant mineral in limestone) to break down the mineral by releasing calcium bicarbonate in solution.

## 2 Factors Affecting Glacial Erosion

There can be great variation in the intensity of glacial erosion depending on the characteristics of individual glaciers. Perhaps the most important factor determining the efficacy of glacial erosion is the basal thermal regime of a glacier (Chapter 2). Where glaciers are cold-based (frozen to their bed) erosion is of very low intensity because most of the processes of erosion already discussed require the presence of meltwater beneath the ice. Cold-based glaciers can cause limited abrasion where ice moves around an obstacle, and dilatation and fracture can occur due to changes in stress on the underlying rock. However, all types of glacial erosion operate far more effectively when glacier ice is warm-based. The presence of meltwater at the base of a glacier allows for basal sliding as a component of glacier movement in addition to internal deformation, and this greatly increases abrasion. A glacier must also be warm-based for regelation to occur, an essential part of the process of glacial plucking.

Other related factors that influence glacial erosion are the ice velocity across the bed, ice thickness, the quantity and quality of rock debris, and characteristics of the bedrock.

- Abrasion is more intense when basal sliding is faster because basal debris is moved at a more rapid rate across the bedrock surface.
- Abrasion is also generally more effective when ice is thicker because there is a greater weight pressing upon the underlying material. However, as downward pressure increases, a point can be reached where increased friction between the abrading rock and the bed causes the abrading rock to stop moving over the surface and to become wedged (or lodged) into the bed. At this point any further increase in ice thickness and pressure will tend to reduce abrasion.
- The quantity of rock debris beneath the glacier is important because with too little a glacier will not have the 'tools' to abrade, but too much will increase frictional resistance and slow basal sliding. Therefore, maximum abrasion rates are reached when there is an intermediate amount of rock debris.
- The rates of abrasion will also be higher if the abrading rocks have sharp edges and are harder than the bedrock, and if there is constant replenishment of rock debris from up-glacier.

Rates of glacial plucking are influenced by many of the same factors that influence abrasion.

- A higher ice velocity will create greater shear stress along the bed, and thick ice with abundant basal meltwater under pressure is important for loosening rock and then entraining it through regelation.
- Glacial plucking is also more effective on bedrock that is highly fractured and jointed.
- The permeability of bedrock influences all of the processes of erosion because it influences the amount of meltwater at the glacier/bedrock interface. Some permeability is helpful for meltwater to exploit joints and fractures within rock, although too much permeability may reduce meltwater pressure at the glacier/bedrock interface thereby reducing basal sliding and mechanical processes of glacial meltwater erosion.

In summary, erosion rates are most intense where glacier ice is warm-based, thick and fast, and the bedrock is relatively weak. Erosion is least intense where glaciers are cold-based and the bedrock is hard.

Areas that have been glaciated previously are eroded more slowly during later glaciations because much of the less resistant material will already have been removed and a relatively efficient network of glacial valleys will already exist to discharge the ice.

## 3 Features and Landforms of Glacial Erosion

Landforms result from the interaction between processes and materials/structure working through time. In the case of landforms of glacial erosion, the processes include abrasion, plucking and the effects of meltwater beneath the ice. Materials/structure refers to the characteristics of the landscape experiencing glaciation and includes bedrock type and structure, as well as the altitude, shape and relief of the land (the topography). Over time, landforms of glaciation become more developed as glaciers continue their work. However, processes do not operate at a constant rate through time, and landforms are continually adjusting to changes in process. After glacial retreat, landforms created by glacial erosion are re-shaped by water, wind and mass movement, and therefore glacial features become less distinct unless later phases of glaciation cause new glaciers to expand over the previously glaciated terrain, reworking the material.

Most present-day landscapes of glacial erosion are the product of not one, but of many successive advances of glacier ice because of the alternation between glacials and interglacials that characterises the Quaternary Period, as explained in Chapter 1. The last glacial ended very recently in geological time (just 11 500 years ago), and therefore mountainous areas that no longer contain glaciers still show the effects of glacial erosion clearly. Hence, landscapes of glacial erosion are certainly not restricted to the areas experiencing active glacial erosion today. For instance, there are presently no glaciers in the British Isles, although the effects of glacial erosion are particularly well displayed in the Highlands of Scotland, in the Lake District (north-west England) and in Snowdonia (north Wales).

There are many different features and landforms produced by glacial erosion, and they can be classified in different ways. For example, classifications can be by the dominant erosional process, by the relative altitude at which they form or by their size range. No single classification is ideal because there is so much overlap between different types, and there are always some features that do not fit categorisation easily. For better appreciating the spatial scale of different types, they are presented in order of size in the sections that follow.

### a) Micro-scale features

Micro-scale features of glacial erosion are measured on a scale of metres or less. **Striations** are scratches on hard bedrock caused by glacial abrasion. The striations are formed as debris is dragged across the bedrock surface by the ice, and therefore they tend to be oriented parallel to the direction that the ice was moving. They are usually only a few millimetres deep but can be several metres long. Where the moving ice encounters a protruding crystal or hard nodule on the bed, the rock on the down-glacier side of the nodule is shielded from abrasion leaving a 'tail' of upstanding rock pointing in the down-flow direction. This feature is termed a **micro-crag and tail** and, together with striations, can be useful for inferring the orientation and direction of past glacier movement. Small fractures, chips and gouges occur in hard bedrock when rock debris beneath a glacier is in intermittent, rather than continuous, contact with the bed. **Chatter marks** are irregular chips and fractures in the rock, whereas **crescentic gouges** have a more regular pattern and are usually concave up-glacier. The photograph on the next page shows an example of crescentic gouges on Torridonian sandstone in north-west Scotland. In addition to scratching and gouging rock, glaciers can 'polish' a rock surface when the material moved across the bedrock is fine, mostly sand and silt-sized, rather than coarse and angular.

Together, these various micro-features can be used to determine the maximum altitude of glacial erosion in a highland area. Glacial abrasion can leave behind striations and chatter marks high up on mountainsides where ice scraped along the sides of its valley. Along some mountain slopes it is possible to identify a change in the surface characteristics of the rock that marks the **trim line**. Below the trim line, there are striations and glacially polished rock, whereas above there is no evidence of glacial abrasion, and instead there are block fields consisting of material weathered by freeze–thaw under periglacial conditions (Chapter 6). Trim lines can be used to reconstruct the height of the former ice surface within a valley (and therefore the glacier's thickness), and can also be used to determine which peaks protruded above the ice as nunataks.

Small depressions and grooves sculpted in hard bedrock are known as **plastically moulded forms** or **p-forms**. They range from potholes to sinuous channels several metres long, and they are believed to form from glacial abrasion being concentrated along certain paths beneath a glacier and/or from meltwater abrasion beneath the ice.

## b) Meso-scale features

Meso-scale refers to features and landforms of intermediate scale, ranging from a metre to a kilometre in size. Streamlined bedrock features are among the most common, forming where glacier ice encounters a hard and resistant bedrock knoll. These features are called **whalebacks** or **streamlined hills**. As the glacier moves over the resistant rocky knoll, abrasion all around the rock causes it to be

Crescentic gouges on Torridonian sandstone, north-west Scotland

smoothed, rounded and streamlined into a shape of least resistance to the flow. The abraded surface usually also shows striations and other micro-scale features. The shape becomes blunt and rounded on the up-glacier side, and tapers like the cross-section of an aeroplane wing in the down-glacier direction. These features are most common in areas of hard bedrock that have been subjected to intense glacial abrasion, for example parts of Scotland, Scandinavia, southern Greenland, Canada and Patagonia.

Unlike streamlined hills, **stoss and lee features** are the product of both abrasion and plucking. Abrasion acts to smooth the up-glacier (stoss) side of a bedrock knoll, while glacial plucking makes the down-glacier (lee) side rough and jagged, producing an asymmetric land-form. Where well developed, the feature is referred to as a **roche moutonnée**, and it can vary in size from several metres to hundreds of metres in length. Classic examples are found in New England in areas of granite hills, where craggy bluffs of over 300 m long and 30 m high have been formed by glacial abrasion and plucking. Similar forms are also found in the Cairngorm mountains of Scotland. The plucking occurs when the pressure beneath the glacier is significantly less on the down-glacier side of the rock obstacle in relation to the up-glacier side. As the glacier moves over the top of the rock obstacle, pressure at the ice/bed interface is lessened, and instead of abrading, the glacier 'plucks' rock off the bed by the process of regelation and entrainment.

**Glacial grooves** and **rock basins** are channels and depressions carved in the bedrock measuring many tens or hundreds of metres in length. They form beneath glaciers where rock is less resistant and therefore more easily excavated by abrasion, plucking and meltwater. As a rock basin deepens, it causes further excavation of the depression by causing ice to have a rotational motion. This increases abrasion on the floor of the basin and plucking (removal of material) where the ice moves up and out of the basin on the down-glacier side. Where meltwater is channelled beneath a glacier, grooves can be widened and deepened into relatively large **subglacial meltwater channels** such as the Gwaun Valley in north Pembrokeshire.

## c) Macro-scale features

On the largest scale, macro-scale features are around a kilometre or greater in size. They are major components of a glacially eroded land-scape, and many of the smaller features of glacial erosion already dis-cussed can be found within them.

Cirque glaciers were introduced in Chapter 1 as glaciers that are confined to an armchair- or bowl-shaped depression, known as a **cirque** (or **corrie**), in relatively high altitude and sheltered locations. Cirques form in the places most favourable for snow accumulation and can become the source of glacier ice that may eventually expand from the mountains into lower altitude valleys during a glacial. In the Northern Hemisphere, cirques most commonly form on the north-east-facing side of mountains where it is both shadier and more sheltered from prevailing westerly winds.

The formation of a cirque is summarised in Figure 10. Once a sheltered area has accumulated snow, a periglacial process called **nivation** begins. Nivation refers to the enlargement of a hollow by a combination of freeze–thaw weathering to loosen the rock, and meltwater from the snow in warmer months to help transport the rock debris away. Once a nivation hollow is established, a positive feedback is set in motion where the existence of the hollow favours additional snow accumulation, which then enhances the process of nivation. Once the snowfield within the hollow has grown large enough to produce glacier ice, then the hollow is further deepened by glacial abrasion as the ice takes on a rotational motion. As the hollow develops into a cirque, the headwall becomes steeper by periglacial freeze–thaw weathering and glacial plucking, and the basin floor becomes deepened by abrasion. The rock lip marks the down-glacier boundary of the developing cirque where the rate of abrasion decreases due to reduced basal pressure. This occurs because the rotation causes ice to thrust in an upward direction on the down-glacier side of the basin. With repeated phases of glaciation, cirques become larger and increasingly bowl shaped, as well as more efficient at collecting snow. Well-developed cirques tend to have a ratio of length to height of about 3:1. When cirques are free of glacier ice, they often contain small lakes known as **tarns**.

The erosion of cirque headwalls backward into the slopes behind can result in other large and impressive erosional features. An **arête** refers to a steep 'knife-edged' ridge produced from the intersection of two cirque headwalls on either side of a slope divide. If erosion causes the ridge separating the two cirques to become lower than the peaks on either side, it forms a saddle-shaped gap known as a **col**. If three or more cirques intersect back to back around the flanks of a mountain, a steep narrow peak is produced between them known as a **pyramidal peak** or **horn**, the most famous example being the Matterhorn on the Swiss–Italian border.

When glacier ice moves through mountain valleys, it straightens, widens and deepens them, transforming what were once stream-formed 'V-shaped' valleys in pre-glacial times into what are often described as 'U-shaped' valleys. However, these **glacial troughs** are more accurately described as parabolic in shape. The action of glaciers significantly modifies pre-existing drainage patterns (as described in Chapter 5) so that the landscape becomes more efficient at discharging ice, and the cross-sectional shape of a glacial trough represents the shape of least resistance to glacier flow.

Glacial troughs can be many kilometres long and hundreds of metres deep. A spectacular example of a glacial trough is the

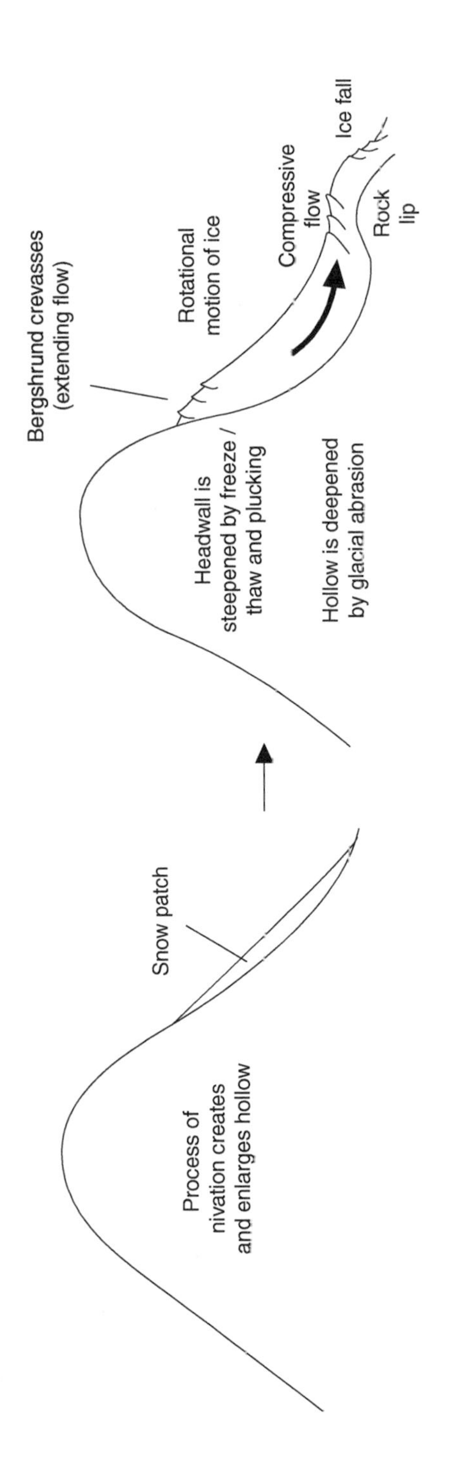

**Figure 10** Schematic diagrams of the formation of a cirque

Yosemite Valley in Yosemite National Park, California. Along their length (the long profile) they contain many smaller features of glacial erosion, for example whalebacks and roche moutonnées, as well as features of deposition (Chapter 4). The floor of a glacial trough is uneven, reflecting both irregularities in the underlying bedrock and differences in the intensity of erosion beneath the glacier. Changes in the slope gradient of the bedrock give rise to areas of extensional flow (ice falls) and areas of compressional flow (Chapter 2). Where flow is compressional, the ice gains a rotational motion (as in a cirque) causing enhanced abrasion and deepening of a rock basin. The development of rock basins is also favoured where bedrock is less resistant to glacial erosion, either because of a weaker rock type or a higher density of jointing. After deglaciation, successive rock basins down a glacial trough, each separated by a lip or 'rock step', may contain lakes. Such a series of small lakes down a glacial valley can resemble a paternoster or string of rosary beads from the air, and they are sometimes referred to as **paternoster lakes**. Longer and deeper rock basins often contain linear lakes termed **ribbon lakes**. With sea-level rise at the end of the last glacial, many coastal glacial troughs were flooded by the sea, of which the **fjords** of Norway and the **sea lochs** of Scotland are good examples.

**Hanging valleys** are another spectacular feature of glacially eroded mountain landscapes occurring where a smaller tributary glacier meets a larger valley glacier. During the glacial phase, the surface ice elevation of the tributary and main valley glaciers is the same, although the rate of erosion beneath the main valley glacier is much greater. This deepens the main valley in relation to the tributary valley, and, once the glaciers have disappeared, the tributary valley can be left 'hanging' hundreds of metres above, often with a waterfall plunging from the hanging valley to the main valley below. Some glacial troughs also contain **truncated spurs**. These represent spurs from the pre-glacial, meandering river valley being gradually cut away or 'truncated' by glacial erosion because glaciers are not able to move around obstacles as easily as streams.

When glaciers expand out of confined mountain valleys they can erode large areas of lower relief producing a landscape of **areal scour**. Such a landscape is composed of many whalebacks, roche moutonnées and rock basins that merge into and out of one another. This type of eroded landscape develops best in areas of hard bedrock that are subjected to extensive warm-based ice, and the structure of the underlying rock has a major effect on the orientation and size of the erosional landforms. In north-west Scotland (for example near Loch Laxford) this open and rugged landscape is referred to as **knock and lochan topography** because of the higher areas of resistant rock (knocks) interspersed with numerous small lakes in the rock basins (lochans). Excellent examples of this type of landscape can also be found on the Canadian Shield, and in west Greenland.

Outside of constricted mountain valleys, glacier ice in the form of an ice sheet can also encounter hills or large rock outcrops, shaping them into giant roche moutonnées measuring many hundreds of metres to several kilometres across. Large **crag and tail** landforms can be produced where glacier ice is forced to flow around a large and resistant rock obstacle. Similar to the formation of micro-crag and tails already discussed, the hard rock obstacle protects less resistant material on the lee side from erosion causing the feature to taper in the down-glacier direction.

## CASE STUDY: GLACIAL EROSION IN THE LAKE DISTRICT

The Lake District possesses many fine examples of landforms of glacial erosion. This upland region in Cumbria, north-west England (Figure 11) owes its existence to ancient volcanism, and much of the central area is made up of hard rock called Borrowdale Volcanics. Since the period of volcanism, over 400 million years of erosion have modified the landscape, and its present form also owes much to the work of repeated Pleistocene glaciation, modifying and deepening the pre-glacial valley system. Many of the lakes (Windermere being the largest) that radiate like spokes outward from the centre of ice accumulation are ribbon lakes that occupy glacially deepened valleys.

The most recent phase of glaciation in the Lake District was during the Loch Lomond Stadial between 12 800 and 11 500 years ago (Chapter 1), a time when glacier retreat at the end of the last glacial was temporarily halted as glaciers re-advanced in upland areas of Britain. However, compared with glaciation during the height of the Devensian, Loch Lomond Stadial glaciation was short-lived, and glaciers in the Lake District remained mostly confined to higher altitude cirques and valleys.

Over 150 glacial cirques have been identified in the Lake District, and some excellent examples are found on the Ullswater side of the Helvellyn Range. The Helvellyn Range runs approximately north–south, including the peak of Helvellyn (950 m altitude) as well as a ridge of high ground over 600 m extending for about 11 km. Most of the cirques of the Helvellyn Range are found on the east side of the ridge, with many facing north-east, this being the orientation most conducive to snow accumulation (Figure 12). During phases of ice-sheet build-up, cirque erosion intensified as local glaciers expanded within their cirques. Owing to the close proximity of many of the Helvellyn cirques, several arêtes have formed. For example, Swirrel Edge and Striding Edge are impressive arêtes that separate the Red Tarn cirque

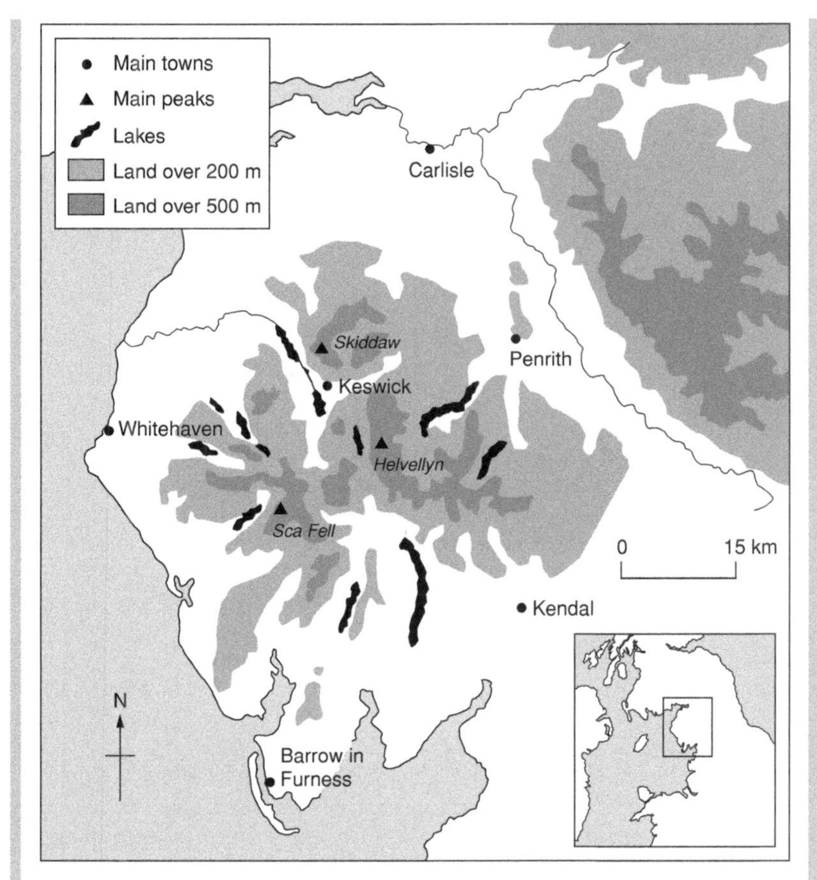


**Figure 11** The Lake District, north-west England

from Brown Cove on the north side and Nethermost Cove on the south side.

The valleys of Grisedale and Glenridding provide excellent examples of glacial troughs. The head of Grisedale contained the largest glacier in the Helvellyn Range during the Loch Lomond Stadial. Fed by ice from at least four different cirques, the glacier extended down valley to an altitude of about 215 m. Above this altitude there are fresh and abundant moraines (Chapter 4). Further down valley beyond the limit of the Loch Lomond Stadial Glacier in Grisedale, moraines from Late Devensian glacier ice are less evident because more time has passed for post-glacial erosion to operate. The Grisedale valley opens out at the southern margin of Ullswater, through which ice was channelled out of the central Lake District in a north-

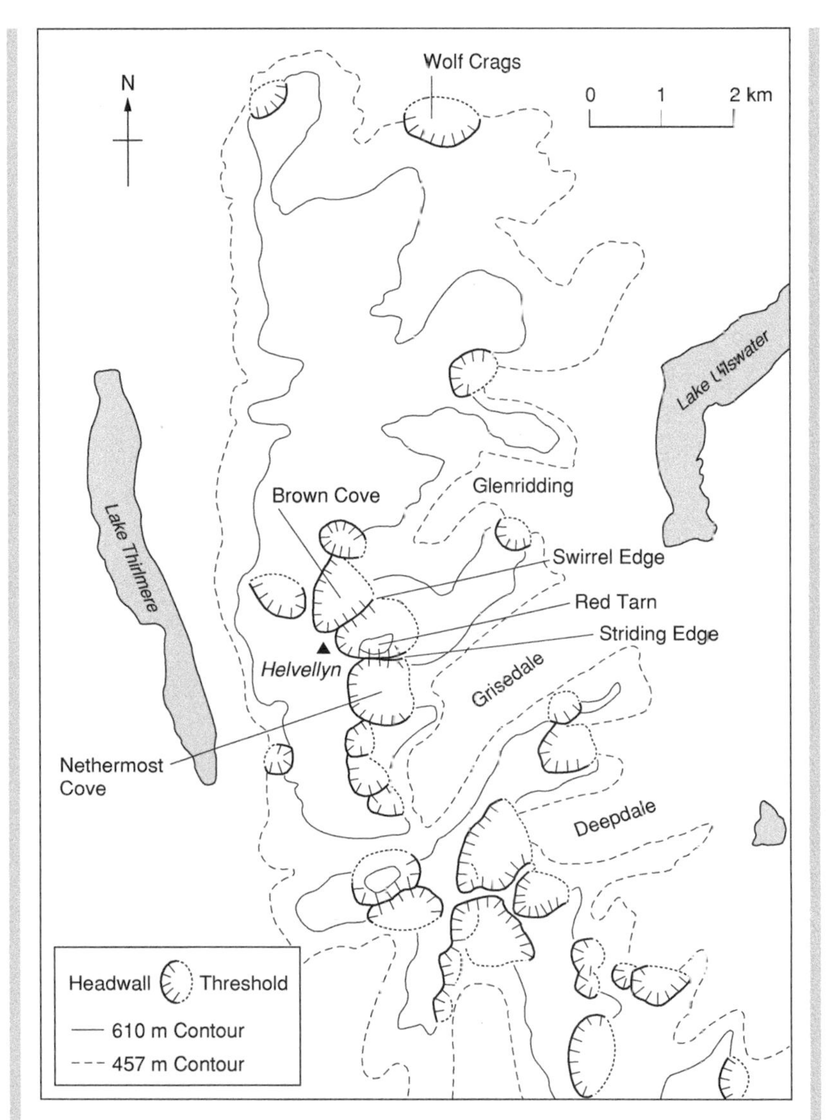


**Figure 12** Cirques, arêtes, glacial troughs and ribbon lakes in the Helvellyn area (modified from Evans, 1997)

easterly direction during the Devensian and earlier glacials. Ullswater itself occupies a glacially deepened trough, but the lake bottom has an irregular long profile because of the presence of sills that provided resistance to glacier ice. Where the lake bends eastward, a roche moutonnée forms a small island in the middle of the lake.

Truncated spurs at Blencathra, Lake District

To the north, the Helvellyn Range is separated from the Blencathra massif by a large glacial trough that trends north-eastward. From near the Wolf Crags cirque at the north-east edge of the Helvellyn Range, there is an excellent view of truncated spurs along the south-facing slopes of Blencathra. The highest points at which the spurs are truncated help indicate the surface altitude of glacier ice that occupied the valley (the trim line). The spurs have become deeply truncated not by one, but by many glaciations over the course of the Pleistocene.

## CASE STUDY: EDINBURGH CASTLE CRAG AND TAIL

A famous example of a glacial crag and tail is the hill on which Edinburgh Castle is built. It is one of many glacially eroded and streamlined features found around Edinburgh. Volcanic activity in the area approximately 325 million years ago (during the Carboniferous Period) has left behind many outcrops of volcanic rock that are more resistant than surrounding sedimentary rock. The castle sits on one such outcrop, known as the Edinburgh Castle crag, that is a solidified magma chamber (volcanic plug) of an ancient volcano. As glacier ice moved east-north-east towards the North Sea, the hard igneous rock making up the crag protected the weaker sedimentary rocks on the eastern (lee) side.

The result is a large crag and tail with a steep up-glacier stoss side and a long gently sloping, tapered lee side that runs for a distance of 1.4 km down to Holyrood Palace. The Royal Mile runs down this slope that is primarily underlain by sediments of the Cementstone Group that pre-date Carboniferous volcanism. On the stoss side, the castle crag stands dramatically at 110 m above a horseshoe-shaped trough scoured out by intense erosion where the ice sheet encountered the crag, moving around and over it. The erosion and streamlining of this large feature has occurred repeatedly over many glacial phases, with the last ice sheet disappearing from the area only around 18 000 years ago.

## Summary Diagram

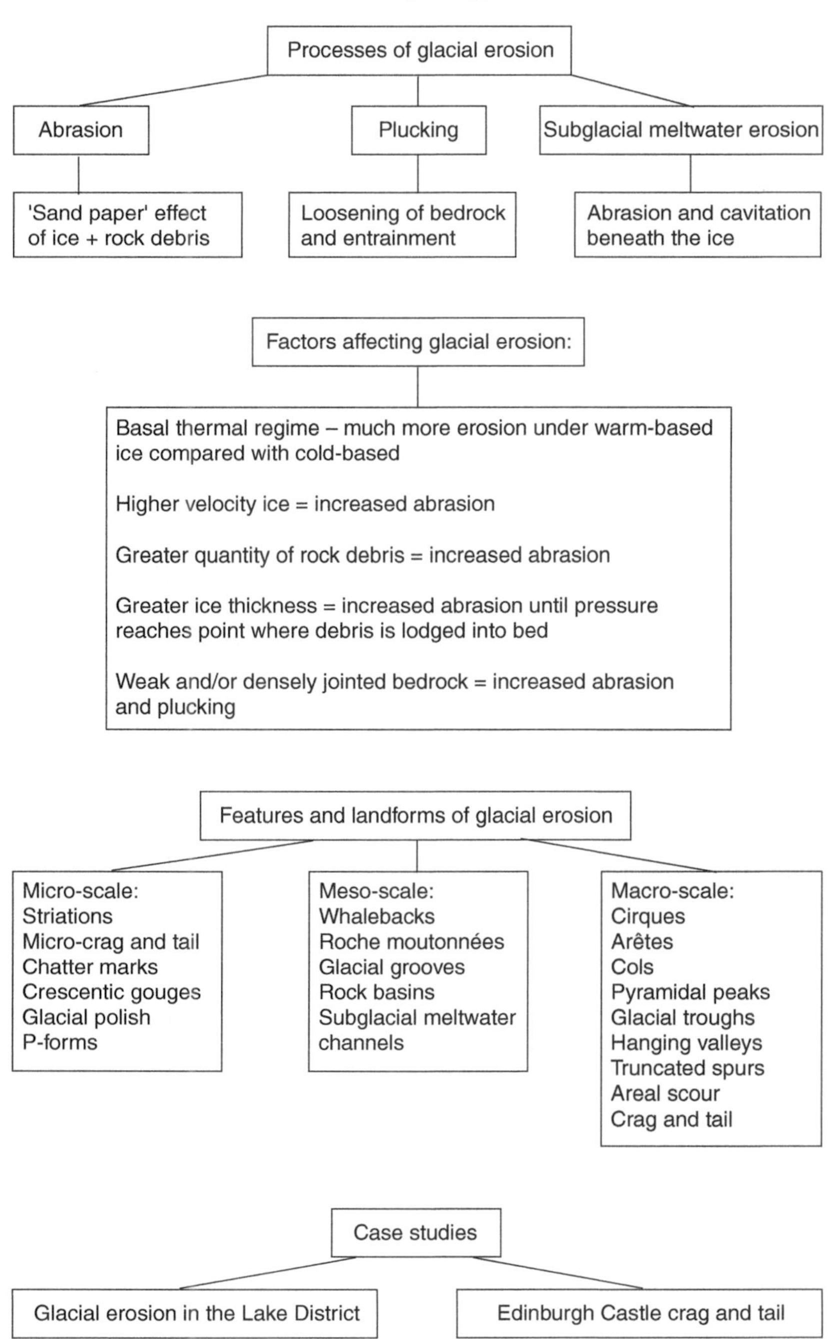

## Questions

**1. a)** Contrast the processes of glacial abrasion and glacial plucking.
**b)** Describe the conditions needed for high rates of glacial erosion.
**2. a)** Describe how features of glacial erosion vary in size.
**b)** Using examples, describe and explain how smaller features of glacial erosion can be found superimposed on larger features.
**3. a)** With reference to an OS map extract of an upland glaciated area (either in the British Isles or elsewhere), locate and describe three landforms of glacial erosion seen on the map.
**b)** Explain how each of these was formed.
**4. a)** Draw an annotated sketch-map of an upland area that you have studied to illustrate features of glacial erosion.
**b)** Describe and explain the effects of glacial erosion in lowland areas.

# 4 Glacial Transport and Deposition

**KEY WORDS**

**Ablation till**: debris deposited by melting of the surrounding ice.
**Erratic**: a rock that has been transported by ice far from its origin and deposited in an area of different rock type.
**Glaciotectonic deformation**: the process by which glacier movement deforms and dislocates underlying bedrock and till.
**Ice-marginal moraines**: moraines formed from glacial deposition along the margins of a glacier.
**Lodgement till**: debris deposited beneath glaciers by the process of lodgement into the bed.
**Moraine**: refers to an accumulation of glacial debris within an active glacier and to the landforms made of till that result from glacial deposition.
**Subglacial moraines**: moraines formed beneath glacier ice.
**Till**: the debris that is deposited directly by glacier ice and not by meltwater. It is characteristically unsorted, non-stratified and unconsolidated.

## 1 Processes of Glacial Transport

Glaciers can transport enormous quantities of rock debris over great distances. In regions previously glaciated this is often shown by the presence of large boulders, known as **glacial erratics**, which are of a different rock type to the bedrock on which they sit. By studying the lithology of glacial erratics it is sometimes possible to match the boulder to its source outcrop, thereby also determining the path that the glacier took. Some glacial erratics were carried many hundreds of kilometres from their source before being deposited by the ice. Boulders weighing up to 16 000 tons were carried by glaciers from the Canadian Rockies and deposited over 300 km away along the plains of southern Alberta.

Rock debris carried by glaciers is derived from both above and below the glacier ice. Along the slopes above a glacier, rocks loosened by mechanical weathering (mainly by freeze–thaw weathering) slide or fall onto the glacier surface. Material may also slowly flow down onto the glacier by the process of gelifluction (described in Chapter 6). Finer material, such as dust and volcanic ash, is carried and deposited over glaciers by wind. Beneath a glacier, rock along the floor and sides of the bed must be plucked and entrained before it can be transported.

Once rock has entered the glacier system and is being transported, it can be classified as either **supraglacial**, **englacial** or **subglacial debris**. Supraglacial debris refers to the material being transported along the surface of the ice. If rock debris from surrounding slopes falls onto the glacier within the ablation zone, it will stay on or near the surface until carried to the snout of the glacier. This is because the ablation zone receives little snowfall and is dominated by melt. If in the accumulation zone, the rock debris will become buried in new snowfall, and over time will be incorporated within the glacier ice as englacial debris. Supraglacial debris can also become englacial by falling through crevasses, particularly where transverse crevasses open up in areas of extensional flow. Subglacial debris refers to the material being transported beneath the glacier along the ice-bedrock interface. It is this basal material that provides the 'tools' for glacial abrasion. Subglacial debris may be transported beneath the ice all the way to the glacier snout, or it may become englacial, or even supraglacial, by moving upwards with the ice along thrust fractures in areas of compressional flow.

Along a glacier, the rock debris being transported becomes concentrated in the ablation zone relative to the accumulation zone, accounting for the much 'dirtier' appearance of the snout at lower elevations. Debris is continually added along the glacier's course, and the melting of ice down-glacier exposes englacial debris on the surface to add to the supraglacial debris. Linear accumulations of debris visible along the surface of a glacier and oriented parallel to its flow are classified as either **lateral** or **medial moraine**. Lateral moraine is the debris along the sides of a glacier, whereas medial moraine is located away from the valley sides. Medial moraine can form either from the confluence of two separate glaciers where two lateral moraines merge to form a central moraine, or from a linear concentration of englacial debris that becomes exposed at the surface by ablation. **Englacial moraine** refers to debris being carried inside the glacier, and **subglacial moraine** refers to accumulations of debris beneath the glacier.

The transportation and eventual deposition of debris by glaciers is as important as glacial erosion for modifying the pre-glacial landscape. By studying the debris left behind after glacier retreat, it is possible to determine not only where the debris came from, but also how it was transported. If carried supraglacially or englacially, the rock will remain largely unaltered from its original state. However, subglacial rock debris is strongly altered by the crushing and grinding that occurs at the ice-bedrock interface. It is more rounded and spherical than debris that has never been in contact with the bed, and it contains a higher percentage of fine particle sizes (**glacial flour**).

## 2 The Nature of Till

The debris deposited directly by glacier ice is called **till**. The term 'drift' has been used for till in older literature, although drift refers

Till deposit at Crummock Water, Lake District

more generally to all types glacial sediments, whether directly deposited by ice or deposited by glacial meltwater. Unlike stream deposits that are well sorted, glacial till is normally unsorted. Fluvial sorting occurs because as a stream loses energy it becomes less able to transport large particle sizes, resulting in a decrease in the particle-size of deposited material with distance downstream. Glaciers, on the other hand, carry a wide range of particle sizes throughout their course, and the resulting deposits contain a mix of larger **clasts** (boulders, cobbles and pebbles) and finer material (sand, silt and clay). For this reason till is sometimes referred to as **boulder-clay**. Till is classified as a **diamicton**, because it is an unsorted, non-stratified and unconsolidated sediment. Non-stratified means that the sediment is not layered and unconsolidated means that the sediment is not compacted or cemented into harder rock. An ancient till deposit that has become consolidated and lithified is known as tillite.

## 3 Processes of Glacial Deposition

The various processes by which glaciers directly deposit material on land are complex and have been classified in a number of ways. However, it is helpful to start by distinguishing between processes of **lodgement**, resulting in **lodgement till**, and processes of **ablation**, resulting in **ablation till**. Till deposits also derive from processes of deformation and flow.

### a) Subglacial lodgement

Lodgement is a process that occurs beneath a glacier when subglacial debris that was being transported becomes stuck or 'lodged' in the bed. It can occur anywhere beneath glacier ice, particularly where the subglacial load of debris is high and the erosive power of the glacier becomes less. While the process is often associated with glaciers in their retreat phase, it also occurs during glacial expansion. Lodgement occurs when the friction between the debris and the bed becomes greater than the drag produced by the ice moving over it.

- An increase in friction beneath the ice that causes lodgement can be due to an increase in ice thickness, in which the weight of overlying ice exerts a greater downward pressure, or due to a decrease in the water pressure beneath the ice.
- The force of the drag beneath the ice is related to the velocity of flow, and therefore a decrease in glacier ice velocity will also favour lodgement.
- The conditions that increase lodgement are the same as those that decrease glacial abrasion.

The lodgement till that results from this process of deposition has a number of diagnostic features. The clasts tend to be relatively rounded and less angular than clasts in other types of till because of the grinding that occurs at the ice/bed interface. The clasts may also have striated surfaces, and the clasts are set within a matrix of clay and silt-sized particles. This often results in a strongly 'bimodal' particle size distribution (the relative proportions of different particle sizes) reflecting the importance of both clasts and the finely ground glacial flour within the deposit. There also tends to be a strong particle fabric (orientation of debris), with elongated clasts oriented with the ice flow direction, and the deposit is well-compacted.

## b) Ablation

The process of ablation, or melt-out, refers to debris being deposited because of the melting away of the ice around it. It can occur subglacially, when geothermal heat melts basal ice causing the deposition of subglacial debris on the bed, or more commonly as supraglacial melt-out along the margins of a glacier when solar radiation melts the ice causing debris on and inside the glacier to be deposited. Glacier ice continually transports debris toward the ablation zone, and the concentration of debris is highest near the snout. At the snout the debris is 'dumped' as the glacier ice that carried the material ablates.

Supraglacial ablation till contrasts strongly with lodgement till. Clasts are largely unaltered from englacial and supraglacial transport and are therefore more angular and less spherical. There is also a lower proportion of fine particle sizes (less glacial flour) than in lodgement till, and the till is less compact.

## c) Deformation and flow

As described in Chapter 2, a glacier can move by subglacial bed deformation if the bedrock material underlying the glacier is relatively weak. When material beneath, or in front, of a glacier is unable to resist the glacier's force, the process of **glaciotectonic deformation** occurs. In addition to folding and faulting of the underlying sediment, this can also result in the incorporation of bed material within the till deposit. The till itself will contain evidence of glaciotectonic deformation in the form of folds and shear planes where different sections of till have been pushed into or over other sections of till. An excellent example of the effect of glaciotectonic processes on till is seen along the Norfolk coast in England near the town of Cromer where large pieces of the underlying chalk bedrock (chalk rafts) are incorporated within the till deposit. The characteristics of deformation till can be summarised as follows. The particle shape and size distribution of the till partly reflect the characteristics of the sediment over which the glacier moved, and the particle fabric tends to be in the direction of the shear stresses (which are not always parallel to the ice flow direction). The till is also well compacted, like lodgement till, and can contain a wide variety of rock types.

Glacial debris can be referred to as 'flow till' if high meltwater content has caused it to creep, slide or flow during deposition. As material becomes saturated with water and the resulting pore water pressure reduces the friction between individual particles, it can behave like a viscous fluid. The speed of flow depends upon the nature of the material and the slope gradient and characteristics of the surface over which it flows. The debris making up flow tills is usually from a supraglacial source, and therefore it is relatively angular and non-spherical. The average particle size is relatively coarse, and the particle fabric shows an orientation reflecting the angle at which it flowed. It is not very compact, and unlike other types of till, it can show some evidence of sorting as a consequence of the flow process.

## 4 Landforms of Glacial Deposition

In a glaciated landscape, landforms of deposition tend to be most abundant at lower altitude while landforms of erosion dominate at higher altitude. Nonetheless, high altitude glacial deposits do occur. Many different landforms of glacial deposition have been recognised, but classification can be made difficult by the presence of transitional types that derived from a combination of the depositional processes described previously. Recognition of different depositional landforms can also be made difficult by soil and vegetation cover and by glacial or fluvial reworking of the deposit. In such cases it may be necessary to dig through the deposit and to sample and analyse the till before it is possible to be certain of the type of depositional form.

The term **moraine** is used to refer to an accumulation of glacial debris whether it is in an active glacier or left behind as a deposit after glacial retreat. Lateral and medial moraines visible at the surface of an active glacier have already been introduced, and there are many other types of moraine that can be recognised in a glaciated landscape after glacier ice has disappeared. In describing the different moraines, it is helpful to separate them into two broad categories – those formed beneath a glacier and those formed along the edges of a glacier.

### a) Subglacially formed moraines

These are moraines that derive from the accumulation of glacial debris beneath glacier ice. This is generally through the process of lodgement described previously, and these moraines are composed mainly of lodgement till. Through the process of lodgement, an ice sheet can deposit till over a wide area. Given enough subglacial debris beneath the glacier, low areas within the pre-glacial landscape can be completely filled in by lodgement till, changing what may once have been a gently undulating landscape into a broad, flat expanse called a **till plain** (also known as a **till sheet** or **ground moraine**). In the British Isles, much of East Anglia is an extensive till plain formed from repeated ice sheet advances. The till averages 30 m in thickness, in some places exceeding 70 m.

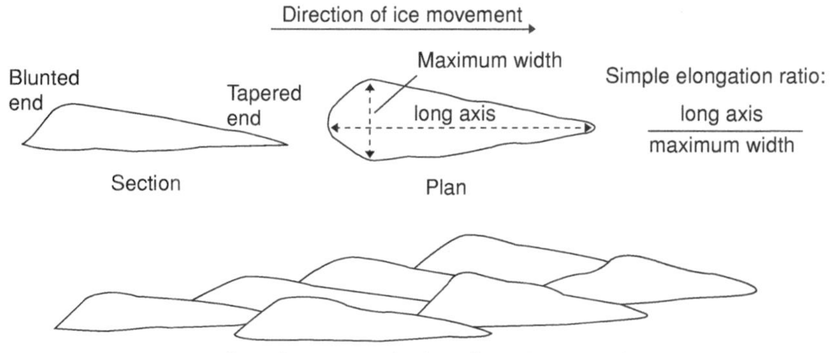


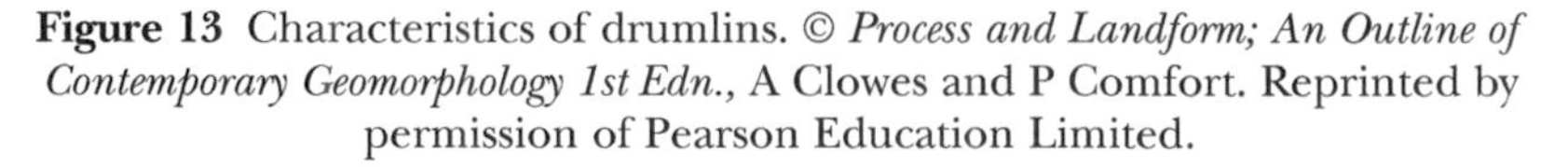


**Figure 13** Characteristics of drumlins. © *Process and Landform; An Outline of Contemporary Geomorphology 1st Edn.*, A Clowes and P Comfort. Reprinted by permission of Pearson Education Limited.

In some places beneath active glacier ice, lodgement till is streamlined into mounds that have the long axis oriented parallel to the direction of ice movement. After glaciation, streamlined mounds of till are called **drumlins** if they have a smooth, elongated, and asymmetric shape (Figure 13). They vary widely in size, ranging from 5 to 50 m high and from 10 to 3000 m long. The ratio of length to maximum width also varies, but is never more than 50. The steeper, blunt end of the drumlin is the up-glacier side, whereas the gently sloping end is the down-glacier side pointing in the direction that the ice was flowing. Viewed from above, the width of the drumlin tapers gradually in the down-glacier direction. The drumlin form represents the shape of least resistance produced by the moving ice as it moulds the subglacial debris. Drumlins are rarely found in isolation and instead are usually found as groups or ‘swarms’, forming what is sometimes called ‘basket of eggs’ topography. They are typically found in lowland areas in relatively close proximity to upland centres of ice dispersal from where there is a high supply of basal debris. Excellent examples of drumlin swarms are found in the Ribble Valley (North Yorkshire) and in the Eden Valley (Cumbria).

The precise way in which drumlins form remains debated, but a number of mechanisms have been proposed. One of the most likely explanations involves drumlin formation in association with subglacial deformation. In this case, a deforming layer of till and bed material is moulded by the glacier in a way that reflects variations in the resistance of the underlying bedrock. Where there is a resistant obstacle within the bed, till is moulded around the obstacle and slower ice flow on its lee side will cause deposition of sediment tapering in the down-glacier direction. Once a drumlin is established beneath a glacier, it will continue to develop because of the way it modifies the flow of ice over and around it. Other explanations invoke the role of meltwater beneath a glacier, suggesting that erosion from large subglacial floods causes scours and irregularities to form in the bed which are subsequently moulded into drumlins.

Streamlined mounds of lodgement till with a length-to-width ratio in excess of 50 are known as **fluted moraine** or **flutes**. These are long and narrow features oriented parallel to the direction that the ice flowed. They are less than 3 m in both height and width, and usually less than 100 m long. If these dimensions are exceeded, the feature is called a **megaflute**. Flutes form when glacier ice encounters resistance from a bedrock obstacle or large boulder or group of boulders. This reduces pressure beneath the ice along a line extending down-glacier from the lee side of the obstacle, causing lodgement till to be deposited as a long ridge (Figure 14).

Till beneath a glacier can also be shaped into moraines that are aligned transverse to the direction of ice flow. These features are known as **Rogen moraine** after the area of moraines around Lake Rogen, Sweden (the term **ribbed moraine** is also used). The ridges that represent individual Rogen moraines are usually between 10 and 30 m high, 150–300 m wide, and can be over 1 km long. The ridge crests are usually spread 100–300 m apart, and they are often slightly concave in the up-glacier direction. In some cases they are formed along large-scale bedrock obstructions that run as a ridge approximately perpendicular to the direction of ice flow. They may also form by accumulation of till where ice flow becomes compressive, thereby reducing basal pressure and encouraging deposition in a long line oriented across the flow direction of the glacier.

## b) Ice-marginal moraines

In contrast with moraines formed beneath the ice, **ice-marginal moraines** derive from glacial deposition along the edges or margins of a glacier. The ice-marginal moraines are named both for where they were deposited in relation to the glacier and for how the material was deposited. For example, it is at the end of a glacier (along the snout) where an **end moraine** forms, although such moraines can also be referred to as **push moraines** if the advancing ice pushes the debris up into a ridge in front of it. Push moraines are a result of the glaciotectonic process, whereby the force of the ice bulldozes sediment in front of it into a ridge. The moraine usually contains supraglacial and subglacial debris, as well as material derived from the rock and sediment in front of the glacier. Along the snouts of active glaciers, small **seasonal push moraines** can be evident produced by the annual advance of the ice margin in winter.

If a moraine ridge along the margin of a glacier has built up simply because of the delivery of debris to the margin without the effect of pushing, it can be called a **dump moraine**. The formation of a large

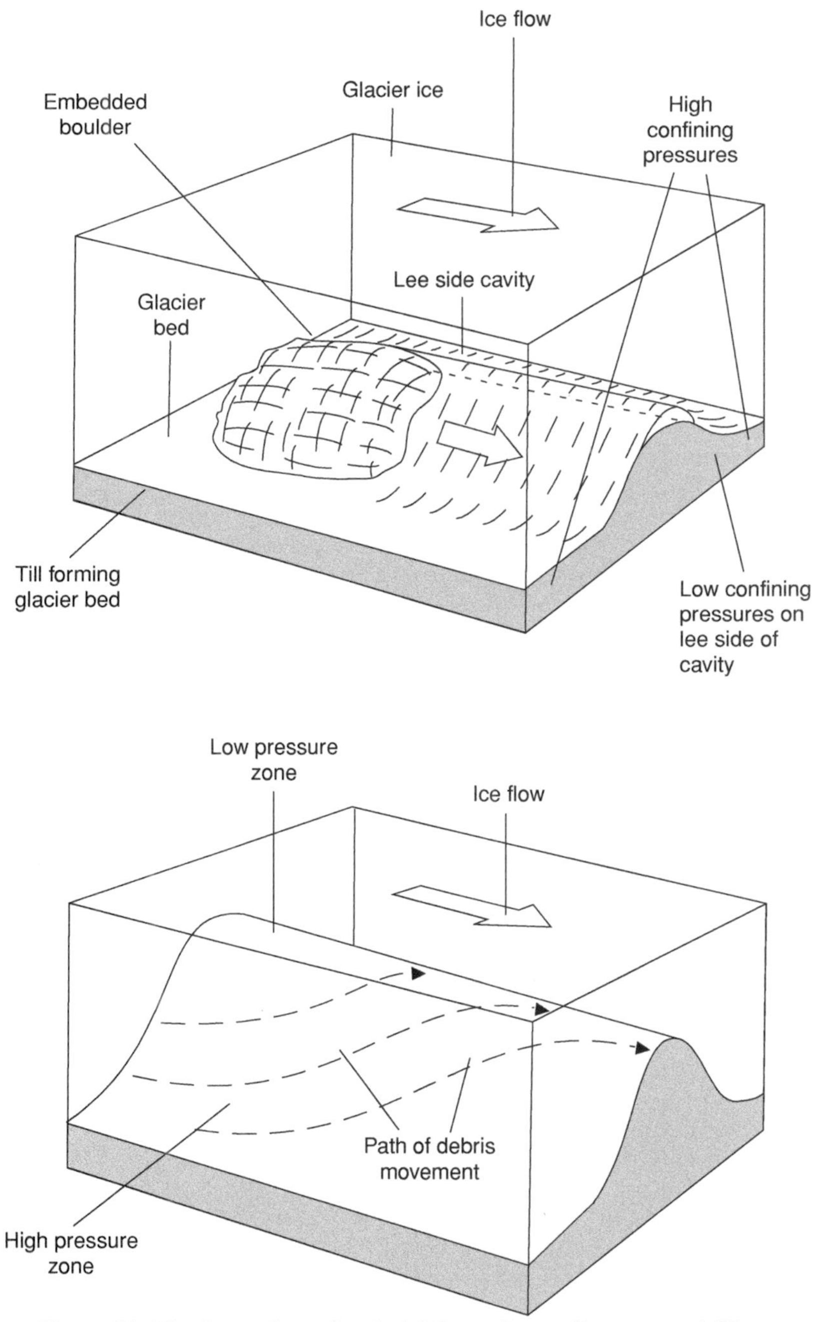


**Figure 14** The formation of a glacial flute. *Source*: Bennett and Glasser (1996).

dump moraine requires that the ice front remain relatively stationary over a long period of time. The size of the moraine will also depend upon the ice velocity and the concentration of debris within the ice. A higher velocity and concentration will result in a greater rate of debris delivery.

At the snout of the glacier, the process of debris 'dumping' can also be responsible for the formation of an end moraine. If the dump moraine has formed along the side of a glacier it is a **lateral moraine** and if it formed down the middle of a glacier (often because of the joining of two lateral moraines where glaciers have merged) it is a **medial moraine**. Lateral moraine remains as a ridge or bench-like deposit along the side of a valley after the glacier has retreated. Medial moraines form ridges along the valley floor, although they tend to be eroded relatively soon after glaciation as post-glacial streams cut new flood plains.

End moraines are called **terminal moraines** when they mark the furthest advance of a glacier, and **recessional moraines** if they mark stages during a glacier's retreat. Terminal moraines can be very large features, sometimes over 100 m high and many kilometres long, particularly where they mark the maximum extent of Late Pleistocene ice sheets across northern Europe and North America. They are formed from a combination of pushing and dumping, and will grow largest where the margin of the glacier remains relatively stationary for a long time. Recessional moraines tend to be smaller than terminal moraines, marking phases during glacial retreat when the ice-margin stayed in one position long enough for an end moraine ridge to build up. Sometimes a re-advance of glacier ice may obliterate previous recessional moraines and create new ones during a later episode of retreat.

In addition to the pushing and dumping processes caused by active glacier ice, ice-marginal moraines can also form as stagnant (or inactive) ice melts away. When there is a high concentration of supraglacial debris in the ablation zone, parts of the glacier margin may become detached from the main body of the glacier. These chunks of detached ice are stationary and melt relatively slowly because a thick cover of debris insulates the ice from surface heating. When debris covers a chunk of stagnant ice it is known as ice-cored moraine. Given a high concentration of debris covering a large area of detached ice, debris will eventually be deposited as a jumble of irregular sized mounds called **hummocky moraine**, also known as 'dead-ice topography'. Some areas of hummocky moraine display a more regular pattern because of the accumulation of debris along former thrust planes.

The different types of subglacial and ice-marginal moraines are summarised in Table 3 according to their shape, orientation with respect to ice flow and mode of formation. Although they form in very different ways, they are often found in close proximity within a glaciated landscape. As discussed in Chapter 5 and illustrated in

**Table 3** Summary classification of morainic forms (simplified from Sugden and John, 1976)

| Linear features | Linear features | Non-linear features |
|---|---|---|
| Parallel to ice flow | Transverse to ice flow | Lacking orientation |
| Subglacial streamlined forms:<br>– Fluted and drumlinised ground moraine<br>– Drumlins<br>– Crag and tail ridges | Subglacial forms:<br>– Rogen or ribbed moraine<br>– De Geer or washboard moraine<br>– Subglacial thrust moraine | Subglacial forms:<br>– Low relief ground moraine (till plain)<br>– Hummocky ground moraine |
| Ice-marginal forms:<br>– Lateral moraine<br>– Medial moraine | Ice front forms: (seasonal/terminal/ recessional)<br>– End moraine (dump moraine)<br>– Push moraine<br>– Ice thrust/shear moraine | Ice surface forms:<br>– Disintegration moraine (hummocky moraine) |

Figure 20, they can also be interspersed with features of glaciofluvial deposition. As glaciers expand and retreat, they rework older glacial deposits into new forms, adding to the complexity of depositional landforms.

## CASE STUDY: VALLEY OF A HUNDRED HILLS, GLEN TORRIDON

One of the most impressive examples of hummocky moraine in the British Isles is located in Glen Torridon, north-west Scotland, approximately 5 km north-east of Torridon Village (Figure 15a). Consisting of numerous small mounds and depressions, this large glacial deposit is known in Scots Gaelic as Coire a' Cheud-chnoic (meaning valley of a hundred hills). From the ground the deposit appears to be a chaotic jumble of till mounds, and it was originally interpreted as a product of ice stagnation near the terminus (snout) of a debris-rich glacier. In valleys lacking clear terminal moraines, areas of hummocky moraine were sometimes

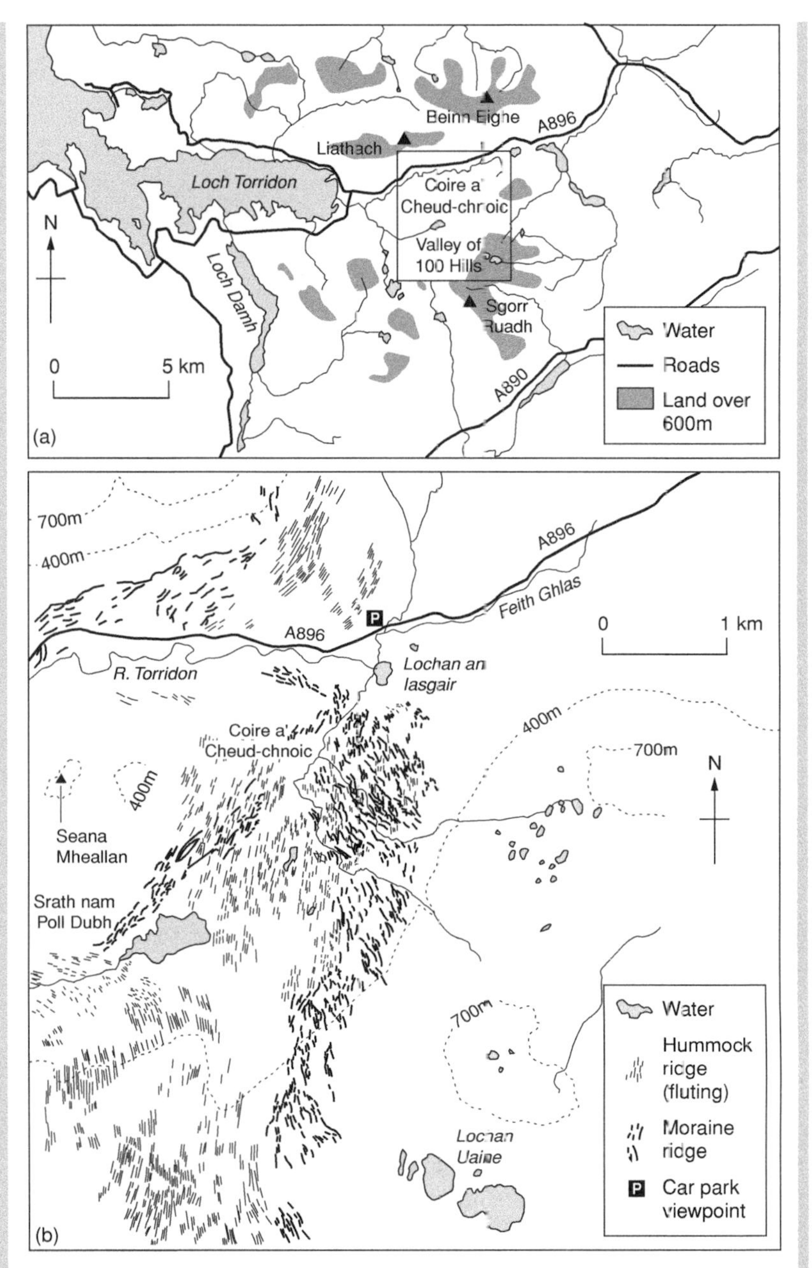


**Figure 15** (a) Location; (b) plan view of Valley of a Hundred Hills. *Source*: Wilson and Evans (2001).

Valley of a Hundred Hills, north-west Scotland

erroneously used to identify the boundaries of Loch Lomond Stadial glaciers within the Scottish Highlands. Valley of a Hundred Hills was believed to mark the maximum eastern extent of a Loch Lomond Stadial glacier that flowed in a north-easterly direction from the slopes of Sgorr Ruadh into Glen Torridon. It was also thought that as the climate warmed, the debris that had built up at the glacier terminus was simply dumped out to form 'dead-ice topography' as the ice beneath experienced rapid ablation.

Later studies revealed that the Valley of a Hundred Hills is a more complex feature than previously thought, and that it cannot be used to infer the glacier's maximum extent. Instead, it now seems certain that ice filled all of Glen Torridon and extended much further to the east during the Loch Lomond Stadial, and that the Valley of a Hundred Hills was formed during a stage in the active retreat of a glacier. From aerial photographs it has been discovered that the hummocky moraine is not as disorderly as it appears from the ground. Instead, most of the till mounds form ridges that are oriented cross-valley (transverse to the former ice flow direction), and in cross-section resemble small push moraines (Figure 15b). It has been hypothesised that this pattern is the result of compressive thrusting within glacier ice near the snout. As glacier ice is thrust up over ice in front, basal debris is also pushed up to the surface along the shear planes where it eventually melts out to form a deposit. As the glacier snout retreats, successive thrust deposits result in a series of till mounds.

Some of the till mounds at Valley of a Hundred Hills are oriented parallel to the former ice flow direction and have therefore been re-classified as fluted moraine. Rather than representing thrusting, these features are believed to have formed when the Loch Lomond Stadial glacier deformed and re-shaped pre-existing till deposited by Late Devensian ice. By studying the Valley of a Hundred Hills it has been shown that areas designated as hummocky moraine can have a more complicated genesis than it first appears and that hummocky deposits can be produced by active glacial retreat as well as by ice-stagnation.

## CASE STUDY: THE ATHABASCA GLACIER

The Athabasca glacier is a valley glacier that flows from the Columbia icefield in the Canadian Rockies. The Columbia icefield itself is the largest body of glacier ice in the Rocky Mountains, covering an area of 325 km$^2$ between the Banff and Jasper National Parks (Figure 16). However, it is a small remnant of the much larger Cordilleran Ice Sheet of western North America that existed during the Wisconsin glacial (the equivalent of the Devensian glacial in the British Isles). The Columbia icefield has an average surface elevation of about 3000 m, and the depth of the ice reaches a maximum of 365 m.

At present the Athabasca glacier is retreating at an average rate of between 1 and 3 m per year. However in recent times its rate of retreat has been even more rapid, and since 1750 it has retreated by nearly 2 km from its maximum 'Little Ice Age' extent (see Chapter 2 case study). This has made the area in front of the glacier an excellent place for the study of moraines. Immediately in front of the snout there are seasonal push moraines. These small ridges are spaced between 2 and 20 m apart, and are typically between 0.7 and 2 m high. They are produced by the seasonal shift of the snout in which there is retreat and dumping of debris in the summer, followed by forward movement of the snout as the glacier gains mass in winter. This causes the debris to be bulldozed up into a ridge that is then left behind as the glacier resumes its retreat in the next summer. As long as the glacier's snout continues to retreat a greater distance in summer than it advances in winter, it leaves behind a series of these moraines that can be used to track glacial retreat year by year.

Further beyond the present-day snout there are recessional moraines marking places where the snout had remained relatively stationary at various periods of time during its retreat (Figure 17). These recessional moraines are larger than the

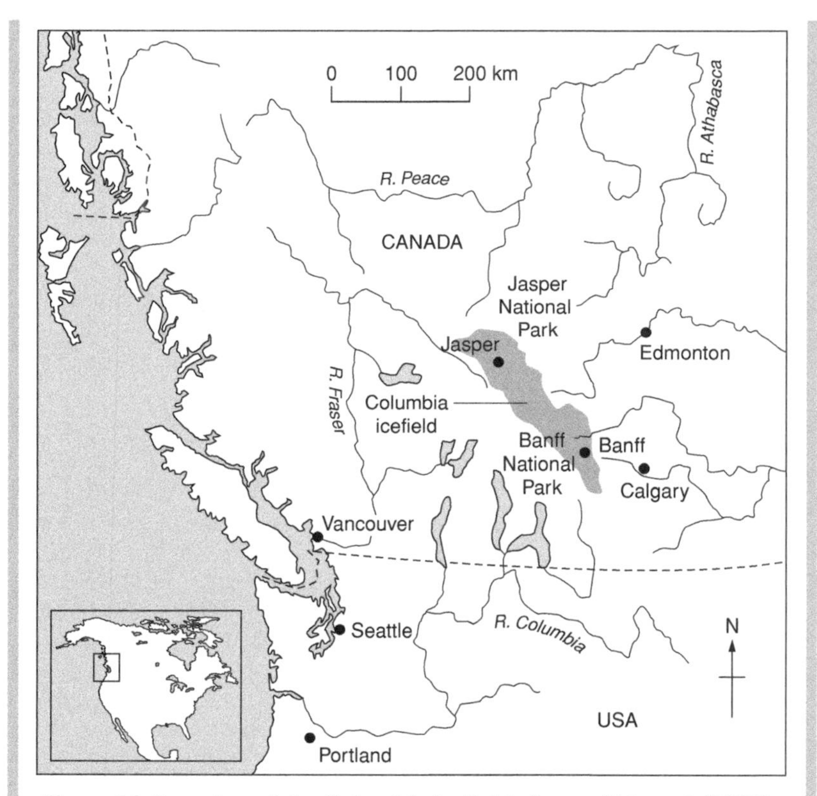


**Figure 16** Location of the Columbia icefield. *Source*: Fishpool (1996).

seasonal push moraines, ranging between 3 and 6 m in height. The furthest of these is the 1870 moraine, and there are also distinct recessional moraines dating from 1938, 1950 and 1960. Knowing the ages of the recessional moraines, the area in front of the Athabasca glacier is an excellent place to study the process of plant succession on deglaciated terrain. This type of succession (glacial foreland succession) is of course much further advanced at the 1870 moraine than inside of the 1960 moraine.

The largest depositional feature is a lateral moraine that extends for 1.5 km along the south-east side of the glacier and stands at a height of 124 m above the valley floor. Its large size results from a high input of rock debris from the slopes above owing to the slopes being composed mainly of well-jointed limestone that is weathered easily by freeze–thaw. This accumulation of debris along the side of the glacier is preventing the ice beneath from melting as rapidly as the rest of the glacier. Over time the lateral moraine will lower as the ice beneath it slowly melts.

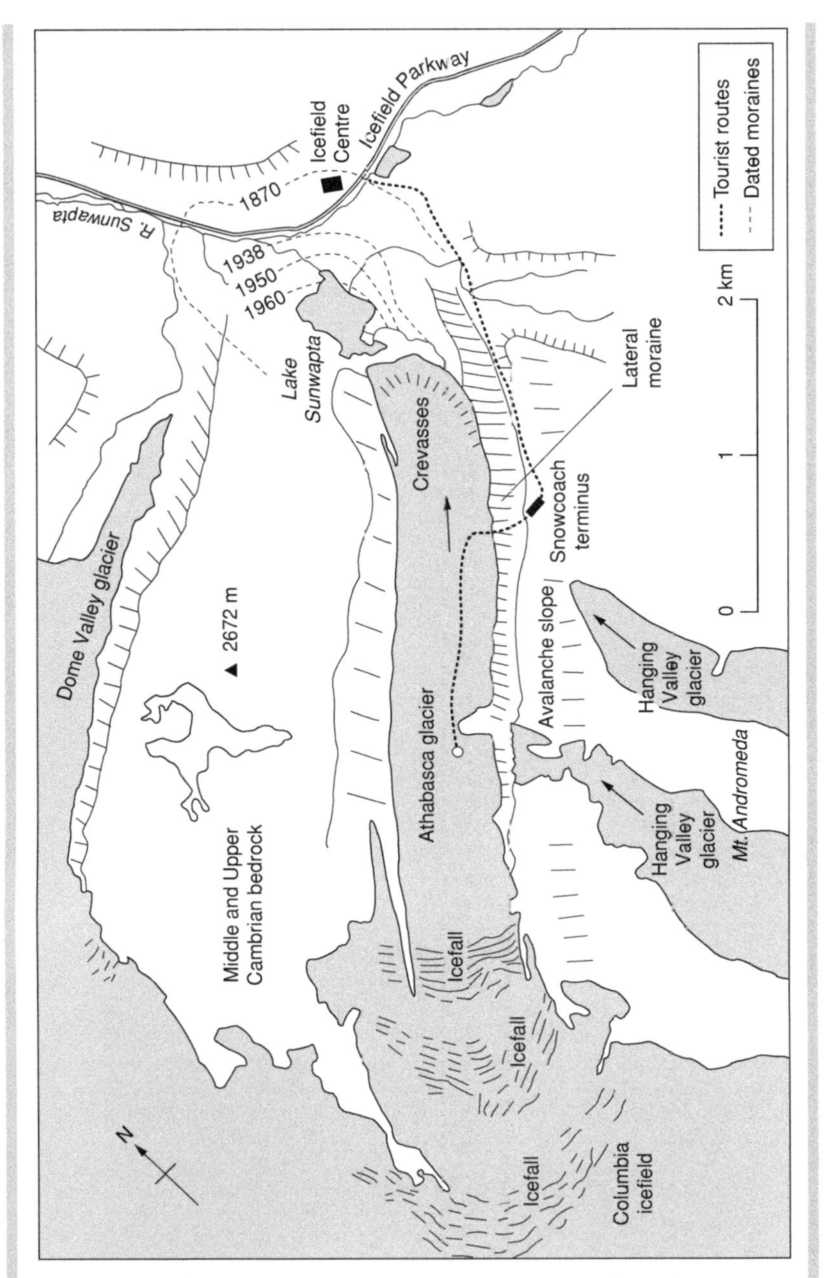


**Figure 17** Plan view of the Athabasca glacier and moraines.
*Source*: Fishpool (1996).

## Summary Diagram

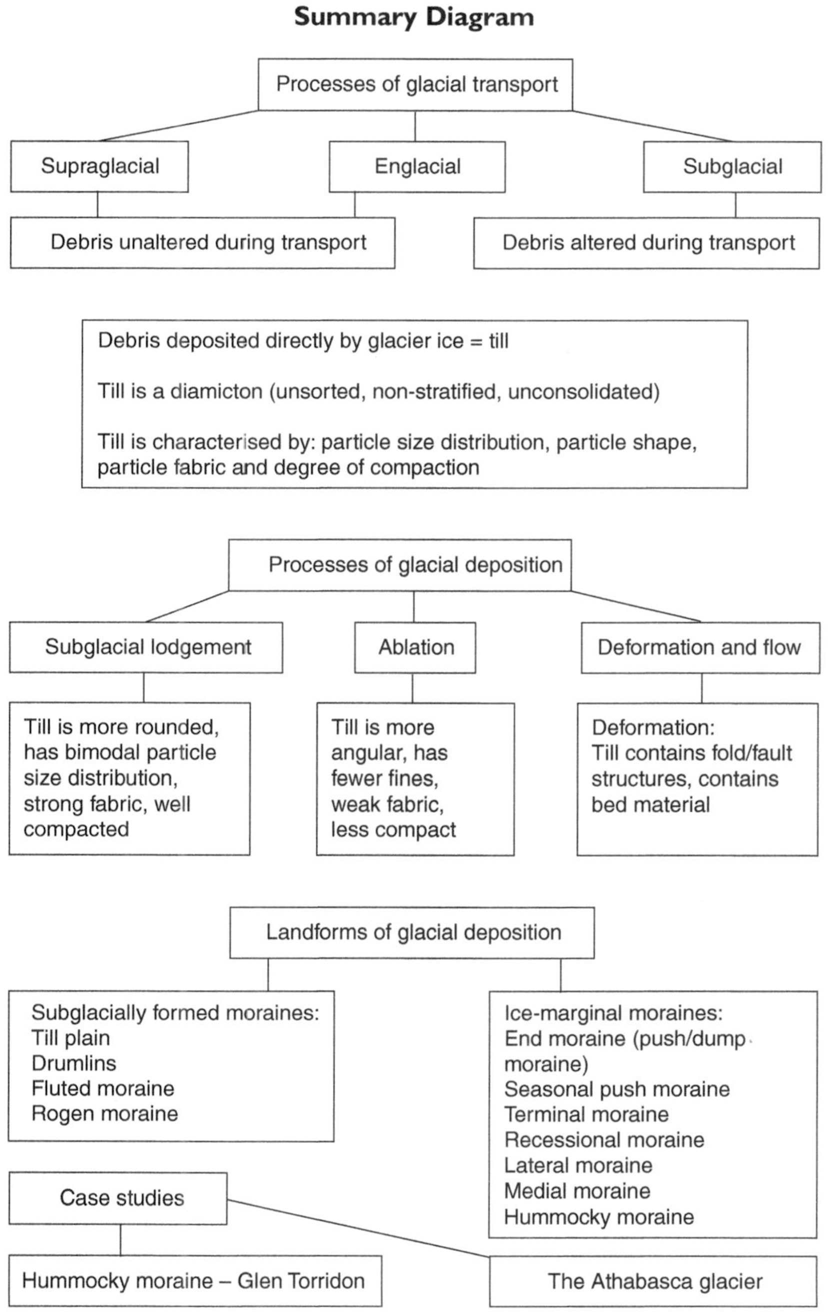

## Questions

**1. a)** Describe the different ways by which material can be transported by a glacier.
**b)** How can the characteristics of deposited material give clues as to how it was transported?

**2. a)** Describe and explain the characteristics of till.
**b)** Compare and contrast lodgement till and ablation till.

**3.** With reference to a specific area, describe different types of ice-marginal moraine.

**4. a)** Using examples, describe different types of sub-glacial moraine and explain their formation.
**b)** Discuss some of the ways by which sub-glacial moraine can be distinguished from ice-marginal moraine.

# 5 Glacial Meltwater and Effects of Glaciers on Drainage

**KEY WORDS**

**Glacial diffluence**: the process by which down-valley flow is impeded forcing glacier ice to breach a pre-existing drainage divide to flow into an adjacent valley.
**Glaciofluvial**: the processes and landforms relating to glacial meltwater.
**Ice-contact glaciofluvial deposit**: material deposited by glacial meltwater at subglacial, englacial, supraglacial or ice-marginal locations. Also known as ice-contact stratified drift.
**Outwash**: material deposited by glacial meltwater streams at or beyond the ice margin.
**Overflow channel**: a steep-sided channel formed by overflow of a proglacial lake.
**Proglacial lake**: a lake formed along the margins of a glacier where meltwater is impounded.
**Stagnant ice**: inactive ice at the margins of, or cut off from, a glacier that melts away as a stationary mass.

## 1 Glacial Meltwater

Meltwater is a key component of the glacial system. As has already been seen, it plays an important role in the motion of glaciers by basal sliding and subglacial bed deformation. It is indirectly responsible for glacial erosion because of its involvement in the processes of glacial abrasion and plucking. More directly, meltwater beneath a glacier has its own erosive effect, causing scouring and grooving of the underlying bedrock. The quantity of meltwater within the glacial system increases toward the snout of a glacier where ablation is at its maximum, and therefore the effects of meltwater are more profound down-glacier than up-glacier. Furthermore, glacial meltwater has far-reaching impacts on the landscape beyond the margins of glacier ice.

There are two main sources of meltwater from glaciers – **surface melting** and **basal melting**. Surface melting accounts for most glacial meltwater, and its contribution peaks in late summer. It is the only source of meltwater from cold-based glaciers that are frozen to their bed. Sometimes surface streams form, running along the top of the ice in the ablation zone, when surface melting produces water more quickly than it can percolate down through the ice. These surface streams are known as **supraglacial channels**. They are typically only a few metres wide and have high velocities because of their smooth sides. Surface streams may plunge down into the ice (becoming **englacial streams**) where there are crevasses, or through cylindrical vertical tunnels called **moulins**. As meltwater moves through the glacier, it may refreeze or contribute to further melting depending on the temperature of the ice inside the glacier.

Basal melting occurs if the temperature of ice at the base of a glacier is at the pressure melting point, as is the case where glaciers are warm-based. The weight of overlying ice combined with frictional heat generated by movement and the accumulation of geothermal heat beneath the glacier can cause basal ice to melt forming a film of meltwater at the glacier/bed interface. This basal meltwater flows under pressure beneath the glacier and, through the exploitation of fractures and weaknesses in the ice, the meltwater is able to excavate **subglacial tunnels** beneath the glacier, eventually exiting at openings along the snout.

## 2 Processes of Glaciofluvial Erosion and Deposition

**Glaciofluvial** action refers to the action of glacial meltwater upon the landscape and the landforms that this produces. Away from a glacier, the fluvial processes by which meltwater streams erode, transport and deposit material are no different from other streams, although the conditions (for example discharge, sediment load and vegetation cover) within which these processes operate are different. Within and beneath a glacier, glaciofluvial processes are quite different from normal fluvial action because of the high pressure and velocity of the flow.

### a) Glaciofluvial erosion

The high pressure and velocity of meltwater beneath glacier ice causes erosion of the underlying bedrock by both abrasion and cavitation, as well as by chemical means, as described in Chapter 3. High amounts of glaciofluvial erosion also occur in front of glaciers in areas where meltwater exiting the ice is channelled. This can result in down cutting through bedrock or through moraines that have been deposited along the ice margin. Such erosion is particularly intense during phases of glacier retreat when increased ablation causes increased meltwater discharge. With increased distance from the ice margin, tributary meltwater streams join to create larger meandering streams that cut floodplains.

### b) Glaciofluvial deposition

When meltwater deposits material subglacially, englacially or supraglacially, the material is referred to as an **ice-contact glaciofluvial**

**deposit** (also known as ice-contact stratified drift). Where glaciofluvial material is deposited at or beyond the ice margin by streams issuing from the glacier, it is known as **outwash**. Unlike glacial till, glaciofluvial deposits have characteristics of normal fluvial deposits, showing evidence of sorting and stratification (layering). In contrast, glacial till is classed as a diamicton, being poorly sorted and non-stratified.

However, ice-contact glaciofluvial deposits tend to have less well-developed features of fluvial deposition than outwash deposits. This is because material in outwash deposits has spent more time being carried by meltwater than material in ice-contact glaciofluvial deposits. Outwash experiences more attrition (particles colliding with other particles) causing clasts to be more rounded, and outwash is better sorted. Three main zones of outwash deposition extend from the front of a glacier, and the characteristics of the outwash change through these zones.

- The **proximal zone** is immediately in front of the glacier close to the snout. It is in this zone where meltwater has the greatest power as it exits the glacier, and, hence, any outwash deposited contains large particle sizes. Outwash in this zone may also be interbedded with layers of till because both direct glacial deposition and meltwater deposition occur along the ice margin.
- The **medial zone** is further from the ice margin, and here meltwater streams tend to form a braided channel pattern because of the high daily and seasonal variability of meltwater discharge and the large quantity of glacial debris being carried in relation to the capacity of the streams. Braided channel systems are characterised by an intricate and mobile network of interweaving channels separated by depositional islands. The particle size of outwash in the medial zone is not as coarse as in the proximal zone, and clasts are more rounded.
- The **distal zone** is the area of meltwater deposition furthest from the ice margin. In the distal zone, the drainage pattern begins to resemble a normal drainage system, characterised by meandering streams, lakes and a broad floodplain where smaller particle sizes are deposited. The outwash is more gently graded, well sorted, and characterised by smaller and more rounded particles than in the proximal or medial zones.

In meltwater lakes along or beyond a glacier margin where water is relatively calm, fine particles that were held in suspension can settle out to deposit thin sedimentary beds, or lamina. These lamina are known as **varves** if they can be resolved into annual bands of sediment reflecting seasonal variation in discharge from a glacier. Varves are characterised by relatively coarse-grained silt or sand at the base of the layer, reflecting more rapid ice melt in summer, grading upwards to finer, dark coloured silt that came out of suspension when the lake's surface, and the streams that fed it, were frozen in winter.

# 3 Glaciofluvial Landforms

## a) Glaciofluvial erosional landforms

Meltwater abrasion by powerful subglacial streams creates a wide range of small moulded p-forms (Chapter 3) that can be found on hard bedrock surfaces after deglaciation. In some cases, subglacial flow can excavate much larger **meltwater channels**. When subglacial meltwater flow occurs under great pressure, and the direction of meltwater flow is controlled by the hydraulic gradient within the glacier rather than by the underlying relief, subglacial meltwater channels can cut across contours. Under these conditions, the resulting channel is said to be **ice-directed**. Subglacial meltwater under pressure can even flow uphill, resulting in meltwater channels where the long profile goes both up and down. The largest example of a sub-glacially cut meltwater channel system in the British Isles is the steep sided Gwaun and Jordanston valley network in north Pembrokeshire, which has a combined length of 20 km and is up to 45 m deep. The Gwaun Valley itself extends over 10 km from its head around the Preseli hills to its outlet at Fishguard Bay.

Glacial meltwater channels are also formed along or in front of an ice margin. The **proglacial streams** that carry meltwater away from a glacier can cut distinct channels or gorges through bedrock or till along a glacier's margin, as seen at the moraine bordering Wolf Crags cirque in the Lake District (see case study in Chapter 3). Large and spectacular gorges have been cut in some places where formerly ice-dammed lakes have drained over steep ridges, or emptied catastrophically as described later in this chapter.

## b) Glaciofluvial depositional landforms

Landforms of glaciofluvial deposition can be formed on, inside and beneath a glacier in 'ice-contact' situations, as well as at or beyond the ice margin as outwash.

The **esker** is the principal ice-contact glaciofluvial landform caused by subglacial meltwater deposition. Eskers are sinuous ridges of relatively coarse sand and gravel deposited by meltwater that flowed through channels or tunnels beneath and inside the ice. Because the deposits are usually laid down in ice-directed channels, they often cut across contour lines, running both up and down hill. Some eskers exist as single ridges while others form part of a complex network of ridges. Eskers can be found at a variety of sizes, with some of the largest measuring many hundreds of kilometres long, hundreds of metres wide and tens of metres high. For example, the Munro Esker in Canada is about 400 km long. Small eskers are less than a kilometre long and approximately 50 m wide and 20 m high. The height and width varies along the length, sometimes causing an esker to resemble a string of beads. Eskers are particularly abundant on glaciated lowlands underlain by hard bedrock, such as in Canada and Scandinavia.

Eskers are thought to form when a subglacial or englacial channel becomes obstructed somewhere along its course by a shifting block of glacier ice. This causes sudden deposition of material along the channel upstream of the blockage, and the material is left behind as a ridge once the glacier has retreated. In the case of englacial (within the ice) deposition, the sinuous accumulation of material is lowered onto the ground as the ice melts. This disturbs the stratification of material contained in the esker, but the sediments in the centre usually preserve horizontal bedding. For an esker to survive from deposition in an englacial position, the glacier ice must become inactive (**stagnant ice**) as it wastes away, otherwise movement of glacier ice would rework the material. Some eskers appear to have formed from a different process whereby a delta of glaciofluvial material extending outward, perpendicular to the ice margin, takes on an elongated form under conditions of rapid glacier retreat.

**Kames** are ice-contact glaciofluvial deposits that derive mainly from supraglacial meltwater deposition concentrated near the glacier snout. The main differences between kames and eskers in terms of their mode of formation and their shape and size are summarised in Figure 18. The formation of kames involves deposition of material in ice-surface depressions and crevasses, as well as along the sides of a glacier between the ice and hillslope. As stagnant ice melts away, the deposit is lowered down to the valley floor forming a mound. Kames are often steep sided and conical, although they come in a variety of shapes and sizes depending upon the dimensions of the ice depression where the material accumulated. The shape of the original fill becomes inverted in the shape of the kame as illustrated in Figure 19. The debris making up the kame usually shows evidence of stratification, although its bedding is often disturbed from the process of subsidence as the ice beneath the debris melts away.

Kames that form a long, relatively continuous bench-like feature along the side of a valley are called **kame terraces**. This type of feature is well exemplified around the shores of Loch Etive in northwest Scotland. Kame terraces are formed when a gap between the valley side and the ice margin is filled with glaciofluvial deposits, leaving behind a terrace as the glacier ice melts. The kame terrace often begins as a delta forming where a supraglacial stream enters an ice-marginal lake between the glacier and valley side. Over time, such a lake may be completely filled with glaciofluvial material that eventually forms the kame terrace as the glacier wastes away.

Regions of extensive kame deposits also contain **kettle holes** and can be described as areas of **kame and kettle topography**. Kettle holes are the opposite of kames, being depressions rather than mounds. They form when a piece of ice buried beneath glaciofluvial material melts. This causes the surface layer of debris in the area

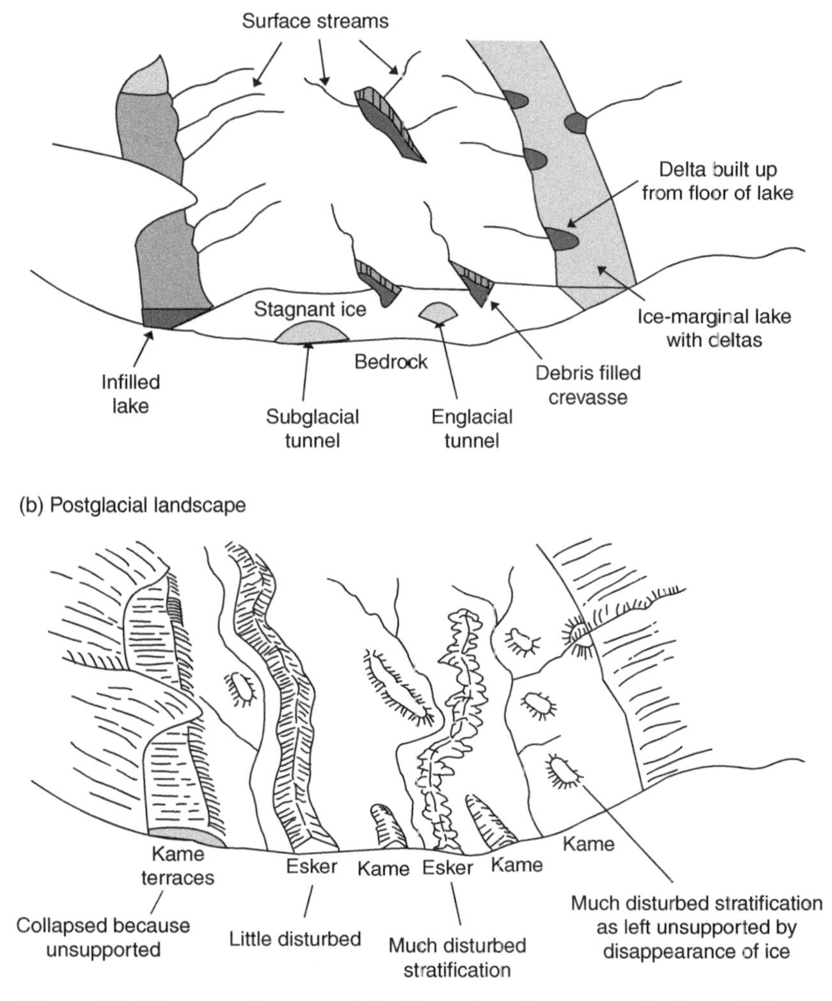


**Figure 18** Schematic diagram of the formation of kames and eskers. © *Process and Landform; An Outline of Contemporary Geomorphology 1st Edn.*, A Clowes and P Comfort. Reprinted by permission of Pearson Education Limited.

above the buried ice to slump relative to the debris around it (Figure 19). Kame and kettle topography is best developed when there is a large amount of glaciofluvial material deposited over the surface of a large area of stagnant ice. After deposition of the glaciofluvial debris, and as the ice melts *in situ*, irregularities that existed in the ice surface become reflected in the pattern of kames and kettles that result from lowering of glaciofluvial material down onto the valley floor. If kettle holes are filled with water they are called **kettle lakes**.

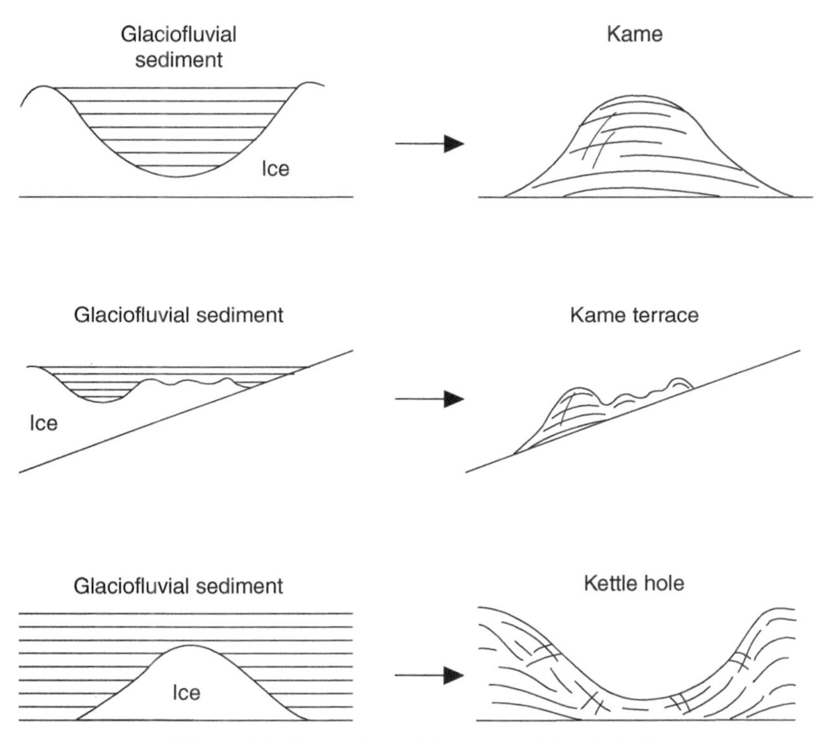


**Figure 19** Formation of kames and kettle holes

Where an englacial meltwater stream exits the snout of a glacier there is a rapid decrease in water pressure and velocity causing the deposition of coarse glaciofluvial material as an **outwash fan** in the proximal zone. Seen in cross-section, the outwash fan slopes gently away in the flow direction but is relatively steep on the side that was in contact with the ice front. For an outwash fan to become large, the glacier's snout must remain relatively stationary for a long period of time.

Several outwash fans can merge away from a glacier to become part of a debris-rich, braided drainage system in the medial zone. As the discharge of meltwater decreases with deglaciation, the broad expanse of glaciofluvial material that was deposited and spread out by the braided stream system is left behind as an **outwash plain** (also known as **sandur**). An outwash plain is a gently sloping surface made up of rounded, sorted and stratified sands and gravel, with the particle size becoming progressively smaller with increased distance from the former glacier margin. They may also contain kettle holes where any surviving pieces of ice during deglaciation were buried by outwash. Outwash plains vary in size, with large examples covering hundreds of

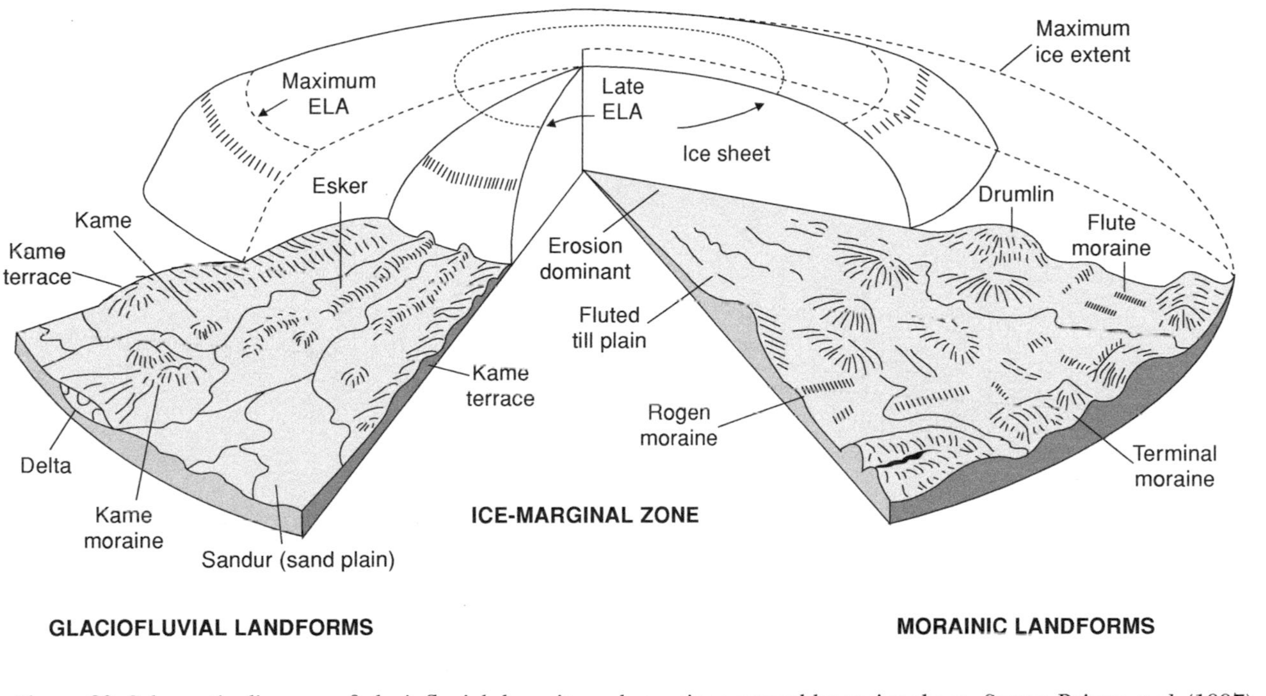


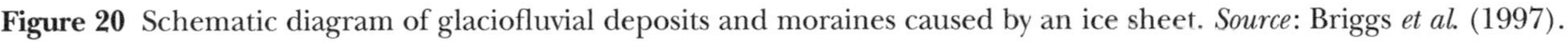
**Figure 20** Schematic diagram of glaciofluvial deposits and moraines caused by an ice sheet. *Source*: Briggs *et al.* (1997).

square kilometres. They tend not to contain clays and silts, because this finer material is transported much further from the glacier by meltwater streams. As a result, the coarse soils that form on outwash plains have little water retention and are nutrient poor. Where an outwash plain forms in an enclosed valley, it is sometimes referred to as **valley train**.

The various types of glaciofluvial deposits and moraines that can be found in a previously glaciated lowland area are summarised in Figure 20. It is important to realise, however, that not all of these types will necessarily be present in a glaciated area. The quantity, combination and eventual form of deposits left behind are influenced by the concentration of debris carried by the glacier ice, by the amount of meltwater and by the style of glacial retreat. For instance, the development of extensive kame and kettle topography requires that much of the ice in the ablation zone becomes detached from the retreating glacier or ice sheet to become stagnant ice, thereby wasting away *in situ* beneath large quantities of glaciofluvial material. In contrast, if deglaciation is relatively rapid and the ice margin retreats back more quickly (lessening the amount of debris accumulating on the ice surface), an outwash plain will be developed in front of the retreating glacial margin with few kames or kettles.

# 4 Effects of Glaciation on Drainage

In addition to producing a remarkable range of erosional and depositional landforms, glaciers and glacial meltwater can also be responsible for major changes in the drainage system. Such changes include the formation of proglacial lakes, the breaching of drainage divides, the diversion of streams and the excavation of deep overflow channels.

## a) Proglacial lakes

**Proglacial lakes** (also known as ice-margin lakes) are formed along the front of glaciers in cases where meltwater exiting a glacier becomes impounded within a depression. Proglacial lakes are usually contained within an area that is blocked by glacier ice and bounded by high ground. They range widely in size and shape depending on unique local characteristics of relief and the position of ice margins. Such lakes are not permanent features, and instead proglacial lakes can lower rapidly as glacier margins retreat, allowing water that was previously dammed by ice to outflow. Depending on the nature of deglaciation, a proglacial lake can either empty completely in one episode of outflow or stabilise at a lower level if the ice margin impounding the lake has retreated but not disappeared.

The dimensions of former proglacial lakes can be inferred from both erosional and depositional features. Given a large enough lake that had a stationary water level for a relatively long time, there can be a well-developed line of erosion running horizontally across a hillside. Such a linear feature cut into a hillside, bench-like in cross-section, is termed a **strandline** and represents a previous lake shoreline. The margins of proglacial lakes can also be reconstructed from the height and position of deltas where meltwater streams deposited outwash as they entered the lake.

In the Western Highlands of Scotland, a proglacial lake that occupied Glen Roy during the Loch Lomond Stadial has left behind a series of impressive strandlines of 6–7 m in width along the slopes on either side of the valley. These strandlines are known as the Parallel Roads of Glen Roy. At its maximum extent the lake was about 16 km long and as much as 200 m deep. It experienced three main episodes of outflow as the glaciers damming the lake retreated, and this has left three principal strandlines. The highest strandline, corresponding to the highest lake level, is at 350 m above sea level, and the middle and lower strandlines are at 325 and 260 m. Similar ice-dammed proglacial lakes also formed in neighbouring Glen Gloy and Glen Spean.

Some of the largest proglacial lakes of the last glacial formed along the margins of ice sheets in North America and Eurasia. For example, the Laurentide ice sheet in North America blocked drainage of the northern Great Plains into Hudson Bay causing the formation of Lake Agassiz, which extended from what is now central Canada into North Dakota and Minnesota in the USA. This massive lake was larger than all of today's Great Lakes combined, being approximately 700 miles long and 200 miles wide and covering an area of around 300 000 km$^2$. As the Laurentide ice sheet retreated, Lake Agassiz eventually drained into Hudson Bay leaving behind the present-day Lake Winnipeg, Lake of the Woods, Lake Winnipegosis and Lake Manitoba as remnants. Further west, glacial Lake Missoula, which covered about 7500 km$^2$ and contained as much water as the present-day Lake Erie and Lake Ontario combined, formed along the margin of the Cordilleran ice sheet.

Large proglacial lakes also formed along the margins of the former British ice sheet. For instance, glacial Lake Harrison formed in the English Midlands during the Wolstonian glacial (the glacial preceding the Devensian) when a lobe of ice blocked the River Avon and meltwater was unable to drain into the sea via the Severn River. The area submerged by glacial Lake Harrison stretched from north-east to south-west including what is now Leicester, Rugby, Coventry, Warwick and Stratford. It was bounded to the north by ice blocking the River Tame and River Soar valleys, and to the east by the Jurassic escarpment of the Cotswolds.

## b) Breaching of drainage divides

In upland areas, glacial erosion can result in the breaching of a pre-glacial drainage divide causing post-glacial streams to be diverted through the breach. This is caused by the process of **glacial diffluence**. If the flow of a glacier down its valley should become impeded by another glacier, the obstructed glacier ice is forced to exit the valley along a different route. Glacier ice may also be forced to find an additional route if it is accumulating more rapidly than it is being discharged down valley. The **diffluent ice** is able to escape from a constricted valley most easily where there is weak rock or a col between surrounding uplands. As the new outflow route becomes established, the col between valleys becomes widened and deepened through glacial erosion into a trough, eventually forming a pass or 'breach' that cuts through what was once a continuous drainage divide.

Many post-glacial streams in areas that were affected by glacial diffluence have acute hairpin bends where they change direction and flow through a breach into an adjacent valley rather than follow the expected path down valley from their source. The number of breaches between adjacent valleys in a glaciated upland region gives a good indication of the severity of glacial erosion. Such breaches are more common in highly eroded upland areas, and the resultant landscape is characterised by a high degree of connectivity between valleys. In contrast, areas less affected by glacial erosion preserve a more dendritic drainage pattern produced by fluvial erosion with a lower degree of valley connectivity. For example, the north and west Scottish Highlands were subject to more intense glacial erosion than areas further south and east, and hence contain many more glacially breached drainage divides.

On a larger scale than glacial diffluence, **glacial transfluence** results in the breaching of a regional divide in many places. The ice cap in north-west Scotland was thickest to the east of the pre-glacial divide that separated streams flowing to the North Sea from those that flowed to the Atlantic. As a result, glacier ice flowed westward across this pre-glacial divide causing it to be extensively breached. This has displaced the regional divide eastward relative to its pre-glacial position, and the upper courses of streams that used to drain eastward now flow to the west through more than 30 glacial breaches.

## c) Diversion of streams

If glacier advance causes a river valley to become blocked by ice, the river is forced to change course. This process of **stream diversion** results in the creation of a new valley and floodplain along the margins of the ice that continues to be occupied by the stream even after glacial retreat. The present course of the River Thames in south-east England is the result of repeated stream diversion by advancing ice during the Anglian glacial. The pre-glacial Thames flowed north of its present course making its way to the North Sea through the Vale of St

Albans and across East Anglia. A lobe of the British ice sheet advanced from the north and blocked this route creating a proglacial lake along the ice margin and diverting the Thames along a more southerly route that is thought to have passed just north of present-day Harrow and Hampstead. Later during the glacial, an advance of ice also blocked this route and forced the Thames further south into its present floodplain. The diverted Thames had a much higher discharge during the Anglian, and later glaciations, than it has today because of the contribution of glacial meltwater. This made it more effective at cutting and widening its new floodplain.

## d) Overflow channels and catastrophic drainage of proglacial lakes

**Overflow channels**, also known as spillways, are among the most impressive features produced by glacial meltwater. They are formed by proglacial lakes overflowing their confines. As water rapidly discharges out of a proglacial lake, intense fluvial erosion occurs along the outflow path often creating a deep gorge that remains long after the proglacial lake has disappeared. For example, the Kirkham Abbey Gorge in North Yorkshire is believed to be the result of overflow from glacial Lake Pickering as it drained southward into the Vale of York. Glacial Lake Pickering formed between the Cleveland Hills to the north and the Yorkshire Wolds to the south when glacier ice blocked outlets to the east and west. It reached a maximum of 68 m above sea level before overflowing southward and cutting the Kirkham Abbey Gorge. This also changed the course of the River Derwent, which used to flow directly into the North Sea but now flows south-westward through the Kirkham Abbey Gorge to eventually join the North Sea via the Humber.

The Channeled Scabland region of Washington State provides the largest-scale example of channels cut by outflow from a proglacial lake. This extensive region in the north-west of the USA contains numerous streamless gorges (also known as **coulees**) that formed during periods of catastrophic flooding from glacial Lake Missoula, which had an enormous volume of about 2000 $km^3$. The lake was dammed on its western side by a lobe of the Cordilleran ice sheet. Against this dam the water reached a depth of about 700 m. During the later stages of the last glacial this ice dam failed repeatedly causing huge amounts of water to flood westward across the Columbia Plateau cutting 100-m deep gorges into the basalt bedrock. It is estimated that at peak flow the discharge from Lake Missoula was $21.3 \times 10^6\ m^3\ s^{-1}$, equivalent to about 100 times the average discharge of the Amazon River. Outflow from Lake Missoula reached speeds of about 65 mph and had tremendous power, producing giant ripples, deep rock basins and wide potholes along the floor of the gorges and carrying boulders with a diameter of up to 10 m.

## CASE STUDY: GRAND TETON AND YELLOWSTONE GLACIATION

Grand Teton and Yellowstone National Parks (north-west USA) contain spectacular examples of glaciation, particularly with respect to the role of meltwater and the effects of glaciers on drainage. Grand Teton is entirely within the state of Wyoming, whereas Yellowstone, just to the north, is mostly contained in Wyoming but also extends into southern Montana and eastern Idaho. The two parks are nearly joined, the southern entrance of Yellowstone being only about 10 km north of the northern entrance of Grand Teton. This close proximity means that the two national parks are often viewed as part of a single 'Greater Yellowstone' ecosystem that also includes surrounding national forests. The glacial history of the two parks is also closely connected.

There is evidence for at least three major episodes of glaciation in the Teton–Yellowstone area during the Pleistocene. The first episode that can be inferred from existing evidence is the Buffalo glacial stage and the second is the Bull Lake glacial stage. Little is known about the extent of the former, but the latter is known to have been the largest of the three episodes. The most recent episode is called the Pinedale glacial stage. It is thought to have begun sometime between 50 000 and 25 000 years ago, and to have ended around 14 000 years ago, therefore being contemporaneous with Devensian glaciation in Britain. After the Pinedale stage, glaciers disappeared completely from the area as precipitation declined and temperatures rose with the onset of the Holocene interglacial. Today about a dozen small glaciers are found above 3000 m in the Teton Range that are believed to have formed during the Little Ice Age cold phase, which began in the 14th century (see Chapter 2 case study).

The reconstructed limits of the Bull Lake and Pinedale stage glaciers are shown in Figure 21. Both glacial stages saw the growth of a regional ice cap centred over Yellowstone and the expansion of alpine glaciers in the Teton Range. This regional ice cap was separate from the much larger Cordilleran ice sheet centred over the Canadian Rockies. Yellowstone was the main centre of ice accumulation in the region because it is mostly a broad plateau with altitudes ranging around 2000 m. At its centre the Yellowstone ice cap had a surface elevation of nearly 3500 m above sea level and was over 1000 m thick. The Teton Range, on the other hand, is relatively narrow and lacking the broad upland area needed to nourish an ice cap. The Bull Lake glacial stage saw the advance of ice southward from Yellowstone filling all of Jackson Hole (the large valley south of Yellowstone

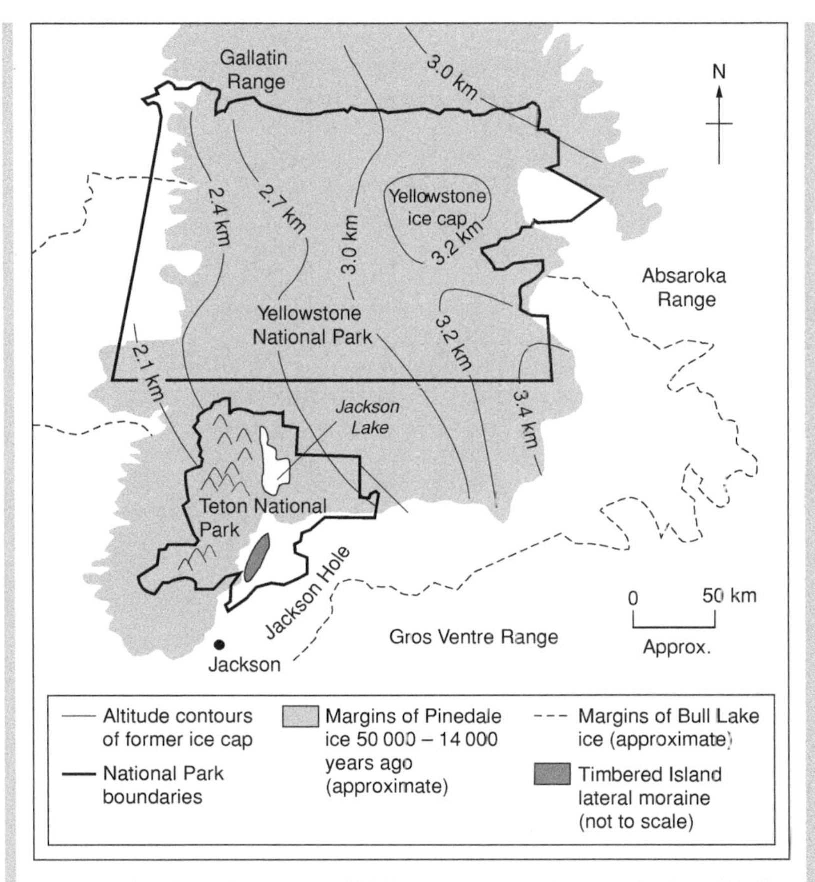


**Figure 21** Grand Teton and Yellowstone area showing limits of Bull Lake and Pinedale stage glaciations

and east of the Tetons), and alpine glaciers flowing down from the Tetons merged with this ice. During the Pinedale glacial stage the Yellowstone ice cap only extended into the northern part of Jackson Hole, and several of the alpine glaciers from the Tetons remained separate, terminating where they exited their valleys along the western margin of Jackson Hole.

Alpine glaciation of the Teton Range, and the convergence of glacier ice from the Tetons and Yellowstone into Jackson Hole, has resulted in some classic examples of glacial erosion, glacial deposition and the effects of meltwater. The jagged Teton Range itself has all the hallmarks of a glacially eroded mountain range, for example cirques, arêtes and glacial troughs. Moraines and glaciofluvial deposits are abundant in Jackson Hole, and are relatively easy to distinguish from each other by virtue of their vegetation cover. The moraines, being made up of till containing clay and silt as well as larger clasts, retain moisture far better than the glaciofluvial deposits of sand and gravel. As a result, moraines can often be identified as tree covered ridges within the otherwise open landscape of Jackson Hole.

With the exception of areas of moraine, the floor of Jackson Hole is mainly a broad outwash plain created as meltwater streams from the Yellowstone ice cap during the Pinedale stage spread glaciofluvial debris across the valley. The sands and gravel of the outwash plain do not retain enough moisture for trees, only supporting relatively drought-tolerant grasses and shrubs like sage brush. For example, a large moraine in Jackson Hole, called Timbered Island, forms a long ridge covered with pine trees aligned north–south that stands out dramatically above the open outwash plain. It is a lateral moraine that formed along the western margin of a lobe of the Yellowstone ice cap that was moving southward through Jackson Hole during the Bull Lake stage glaciation.

Jackson Hole also provides a good example of how glacial erosion, glacial deposition and meltwater action interact to alter drainage patterns. Jackson Lake, the largest lake in the valley, was affected substantially by glacial erosion. The lake existed in pre-glacial times because of the downward tilt of the valley where it meets the east slope of the Tetons, creating a natural lake basin. Glaciers moving down from Yellowstone along the eastern side of the Tetons simply exploited the pre-existing trough to deepen and widen Jackson Lake. A terminal moraine left from the advance of Pinedale ice forms the southern margin of the present-day lake.

The smaller lakes along the foot of the Tetons to the south of Jackson Lake also owe their present form to glacial erosion and damming by terminal moraines. For example, Jenny Lake occupies a basin that was deepened by the Cascade Canyon glacier as it extended into Jackson Hole from its source in the Tetons (Figure 22). During the Pinedale stage the glacier also deposited a broad terminal moraine, which subsequently dammed meltwater streams draining Cascade Canyon to create the lake.

The course of the Snake River through Jackson Hole and the present form of the floodplain have also been greatly influenced by glaciation. The ancestral Snake River once flowed from the south end of Jackson Lake, but it now exits the eastern end because of the terminal moraine that Pinedale ice deposited at the south end of the lake. From Jackson Lake, the Snake meanders across Jackson Hole southward, eventually flowing westward into Idaho. Near Jackson Lake it flows across outwash that was

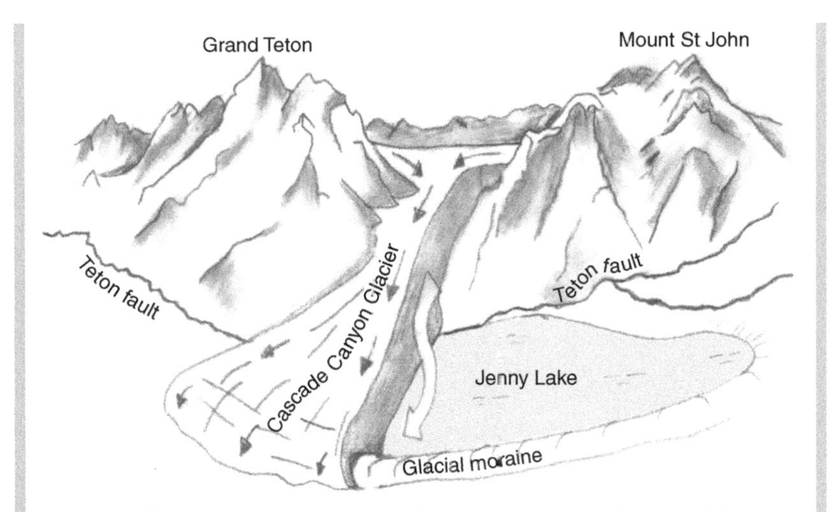


**Figure 22** Formation of Jenny Lake by the Cascade Canyon Glacier. *Source*: Smith and Siegel (2000).

deposited around the margin of Pinedale ice. This area features the 'kame and kettle topography' characteristic of places where glaciofluvial material is deposited along a stagnant ice margin. Further south, the kettles disappear and the outwash plain becomes smoother.

Across Jackson Hole the Snake River has cut impressive terraces into the outwash plain. These step-like benches rising above the present Snake River channel represent former floodplains that existed when the river was at higher levels. The release of meltwater during deglaciation is rarely a gradual process. Instead, there are phases of massive discharge, for example when large glacially dammed meltwater lakes burst, interspersed with periods of relatively low discharge when glacial melting slows or when water is held back in proglacial lakes. During Pinedale deglaciation, periods of high meltwater discharge greatly increased the erosive power of the Snake River causing it to cut more deeply into the outwash plain leaving its former floodplain behind at a higher level. A new floodplain at a lower elevation was then established and broadened in the period of lower discharge until a new surge of meltwater caused the Snake to erode even deeper. It is the alternation of these phases of higher and lower meltwater discharge during deglaciation that has formed the sequence of terraces seen along the Snake.

The Teton Range and Snake River terraces

## Summary Diagram

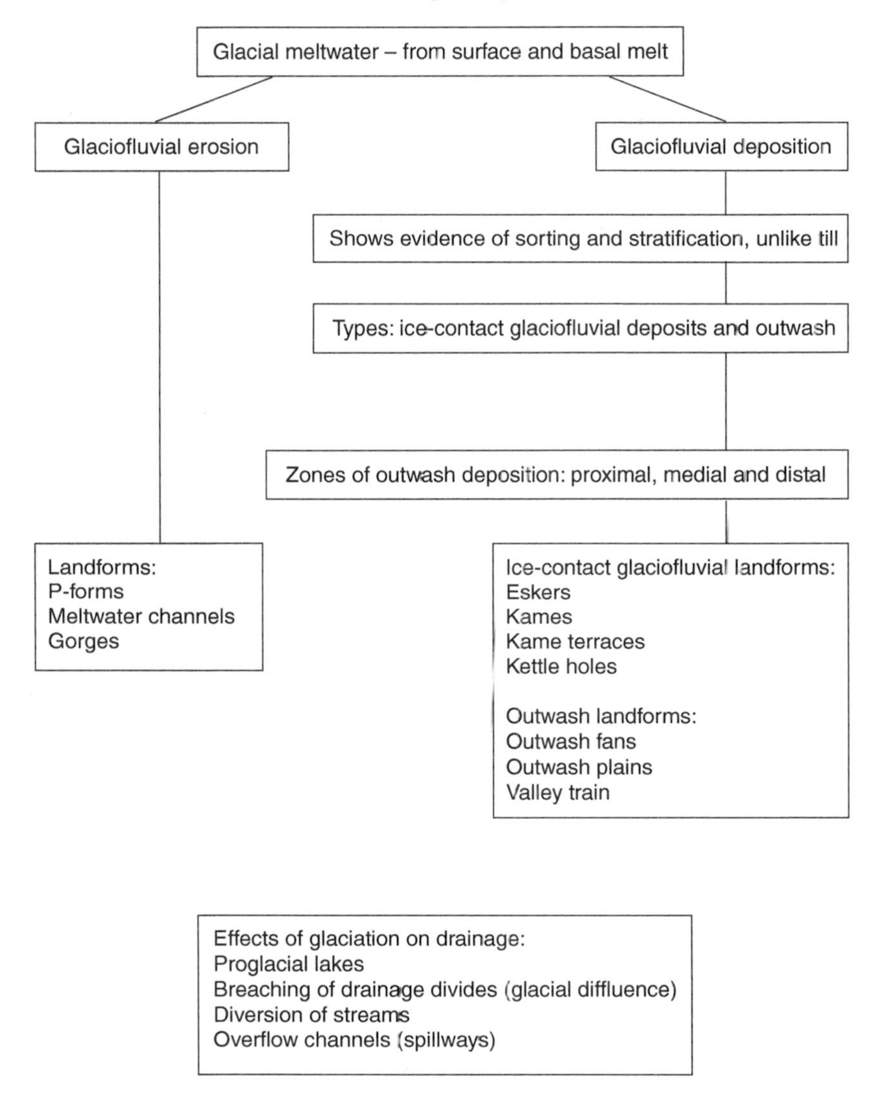


Effects of glaciation on drainage:
Proglacial lakes
Breaching of drainage divides (glacial diffluence)
Diversion of streams
Overflow channels (spillways)

Case study: Grand Teton and Yellowstone Glaciation

## Questions

1. **a)** Summarise the different ways by which meltwater moves through a glacier.
   **b)** Describe how outwash can be distinguished from ice-contact glaciofluvial deposits.
2. **a)** Describe the landforms derived from ice-contact glaciofluvial deposits and explain their formation.
   **b)** Describe the landforms derived from outwash deposits and explain their formation.
   **c)** Why is it often difficult to distinguish between different landforms of glaciofluvial deposition?
3. Using examples, discuss how glacial meltwater produces a range of erosional and depositional landforms.
4. **a)** What evidence in the landscape points to the past existence of a proglacial lake?
   **b)** Explain the process of glacial diffluence and its effect on pre-glacial drainage divides.
5. With reference to specific areas, describe how glaciation can alter drainage patterns.

# 6 Periglacial Processes and Landscapes

**KEY WORDS**

**Active layer**: the top layer of the ground above permafrost that experiences seasonal freezing and thawing.
**Frost action**: the mechanical weathering process caused by repeated freezing and thawing of water in pores, fissures and cracks that forces rock or other materials apart.
**Frost creep**: downslope movement of particles caused by frost heave perpendicular to the slope followed by thawing and vertical lowering.
**Frost heave**: the predominantly upward movement of soil due to expansion as water freezes within it.
**Gelifluction**: the specific type of solifluction (soil flow) that occurs in areas underlain by permafrost.
**Permafrost**: soil and rock in which temperatures remain below 0°C for at least 2 consecutive years.
**Thermokarst**: surface depressions produced by thawing of ground ice in periglacial areas.
**Tundra**: vast treeless plains located poleward of the coniferous forest belt, and dominated by mosses, lichens grasses, sedges and dwarf shrubs.

## 1 Periglacial Environments

The term **periglacial** has long been used to refer to the climatic conditions and landscape that characterised areas near to the margins of glacier ice during the Pleistocene. However, the term is now used more broadly to refer to non-glacial cold-climate processes and landforms regardless of proximity to glacier ice. Periglacial areas therefore encompass a wide range of different environments today, ranging from polar regions bordering existing glaciers to higher altitude areas in mid and low latitudes, which may or may not be near glaciers. Periglacial environments are often, but not exclusively, associated with extensive areas of open, treeless vegetation known as **tundra**. The high-latitude tundra zone begins where the average temperature of the warmest month falls below 10°C, correlating with the poleward limit of coniferous forest. At lower latitudes a similar environment termed **alpine tundra** occurs in mountains above the timberline.

In general, periglacial climates are characterised by long, cold winters during which temperature rarely rises above 0°C. The mean temperature of the coldest month is less than –3°C, and mean annual precipitation is less than 1000 mm. In other respects, periglacial climates can vary considerably, for example in the maximum temperature

**Table 4** Classification of periglacial climates. *Source*: Based on the classification presented by Washburn, A.L. 1979: *Geocryology*. Edward Arnold, London, pp. 7–8.

| Type | Climate description |
|---|---|
| Polar lowlands | Mean temperature of coldest month < –3°C, characterized by ice caps, bare rock surfaces and tundra vegetation. |
| Subpolar lowlands | Mean temperature of coldest month < –3°C and warmest month > 10°C. Taiga type of vegetation. The 10°C isotherm for the warmest month roughly coincides with the tree-line in the Northern Hemisphere. |
| Mid-latitude lowlands | Mean temperature of the coldest month < –3°C but mean temperature > 10°C for at least 4 months per year. |
| Highlands | Climate influenced by altitude as well as latitude. Considerable variability over short distance depending on the slope aspect. Diurnal temperature ranges tend to be large. |

reached in summer and in the nature and distribution of precipitation through the year. While the length and average temperature of the summer can vary, there must be enough warmth to cause ice to melt, because repeated freezing and thawing (or **frost action**) over long periods of time is an essential aspect of all periglacial environments. Ground must also be seasonally free of snow or ice cover. Table 4 summarises the four main types of periglacial climate.

These cold-climate conditions give rise to a variety of processes, known collectively as **periglacial processes** (or periglaciation), which create distinctive landscape features. Some of these processes are unique to periglacial environments, while others (such as frost action) occur elsewhere, but with less frequency and intensity. Approximately 20% of the Earth's land area experiences periglacial conditions today, mostly in the Northern Hemisphere. Periglacial environments are most extensive across Siberia, Alaska, northern Canada and northern Scandinavia, but also occur in mountainous regions throughout the world. The distribution of periglacial environments has shifted dramatically over the Quaternary Period in tandem with the advance and retreat of continental ice sheets over multiple glacial/interglacial cycles, as described in Chapter 1. During the coldest period of the last glacial, periglacial conditions and tundra reached much lower latitudes than today, for example extending across Europe as far as southern France, northern Italy and the Balkans.

## 2 Permafrost

Because of the long, cold winters and short summers, periglacial climates are conducive to the development of **permafrost**. Permafrost is often taken to mean permanently frozen ground, however it can be defined more precisely as the thermal condition in soil and rock where temperatures do not rise above 0°C in the summer months for at least 2 consecutive years. Most water within the permafrost zone of soil and rock will therefore remain frozen throughout the year as **ground ice**. Ground ice takes a variety of forms including frozen pore water, veins through fractures in sediment, and lenses of ice within rock and soil cavities. However, when conditions underground create relatively high pressures and high concentrations of dissolved salts, the freezing point of water can be lowered enough for some groundwater within the permafrost to remain unfrozen.

**Continuous permafrost** forms in the coldest areas of the world where mean annual air temperatures are below –6°C. In parts of northern Canada and Siberia, continuous permafrost extends downward from the ground surface to depths of 500 m or more, and a maximum depth of 1500 m is reached near the upper reaches of the Markha River in Siberia. Yet within the continuous permafrost zone, permafrost may be absent beneath deep lakes that moderate the temperature of underlying ground. Where mean annual air temperatures are between –6 and –1°C, the permafrost is termed **discontinuous** because it is more fragmented and thinner than continuous permafrost. At the margins of this zone there is **sporadic permafrost**, which is very fragmented and sometimes only a few metres thick. It is found where local micro-climatic conditions keep ground temperatures relatively low, such as on the shadier north-facing side of a hill or beneath peat, which insulates the ground from summer thawing.

While periglacial environments often contain permafrost, the distribution of periglacial activity and permafrost is not identical. Estimates of the present distribution of permafrost range between 20 and 25% of the Earth's land area, covering a greater area than the zone of periglaciation. This is because there are extensive areas of permafrost beneath parts of the ice sheets and polar seas where freeze–thaw cycles do not take place. There are also areas of periglacial activity involving intense frost action existing beyond the zone of permafrost. Nonetheless, like the distribution of periglacial environments, the majority of permafrost is in the Northern Hemisphere. As shown in Figure 23, continuous permafrost is mainly located within the Arctic Circle. Russia contains the largest area of permafrost followed by Canada and Alaska. Almost half of Canada and 80% of Alaska contain various forms of permafrost.

The freezing of the ground to produce permafrost begins at the surface and extends downward in winter. The maximum depth of permafrost formation is affected by three key factors – the energy balance

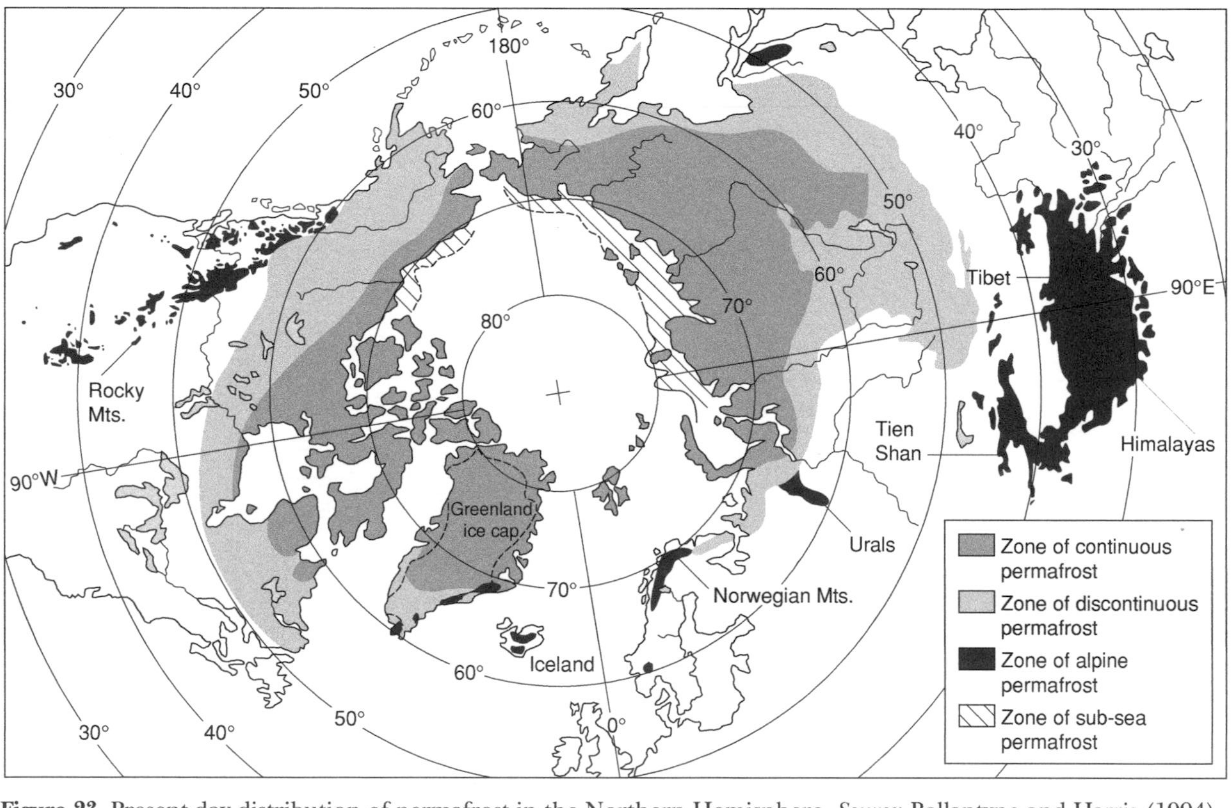

**Figure 23** Present-day distribution of permafrost in the Northern Hemisphere. *Source*: Ballantyne and Harris (1994).

at the surface, the thermal characteristics of subsurface material and geothermal heat flow.

- The surface energy balance refers to the balance between incoming solar radiation and outgoing terrestrial radiation.
- The characteristics of subsurface material are important in relation to the rates at which different materials conduct energy, thereby gaining or losing heat.
- Permafrost will become thickest where the surface energy balance is at its most negative (causing intensely cold winters) and where subsurface material is most conductive of heat.
- Geothermal heat flow ultimately limits the thickness of permafrost because it causes rock temperature to increase with increasing depth. However, a more negative energy balance at the surface means that permafrost can extend to a greater depth before geothermal heat flow is able to raise the underground temperature above 0°C.

In the summer the energy balance at the surface is positive as the daily input of solar energy exceeds the output of terrestrial long wave radiation. This causes snow and ice overlying the ground to melt away, and heat is conducted (transferred) downward from the surface into the ground. This produces a seasonally unfrozen zone above the permafrost called the **active layer**, which reaches its maximum depth at the end of the summer. Depending on the length and intensity of summer heating, the active layer may extend downward by as little as a few millimetres to as much as 3 m. The maximum depth reached by the active layer determines the height of the **permafrost table**, which is the upper boundary of the permafrost. As the energy balance becomes increasingly negative through autumn and winter, freezing occurs from the surface downward through the active layer. However, the rate of downward freezing is not equal throughout the active layer, and pockets of unfrozen ground called **taliks** may survive for many months between the seasonally frozen surface and the underlying permafrost. Taliks can also be found within and below the permafrost.

## 3 Active Layer Processes and Related Features

Repeated cycles of freezing and thawing in the active layer are essential for the operation of many periglacial processes, which in turn produce distinctive features both within the ground and on the surface.

### a) Cryoturbation

As water freezes its volume expands by about 9%. Repeated freezing and thawing within the active layer therefore causes expansion and contraction of the surrounding material beneath the ground. The resultant breaking, churning, and mixing of soil, rock and sediment is known as **cryoturbation**. This can alter beds of sediment that were originally laid down horizontally by twisting and contorting them into folded formations called **involutions**. Fine-grained, unconsolidated sediment is most susceptible to this process due to a relatively high pore water content that makes it slow to freeze. This allows the fine material to deform as surrounding coarser material freezes and exerts pressure.

## b) Segregated ice

As already mentioned, the rate of freezing down through the active layer is unequal, and this results in pockets of ice forming more quickly within some parts of the sediment. Once started, an ice pocket becomes enlarged into an ice mass (**segregated ice**) as surrounding water is drawn to it and freezes onto it. Masses of segregated ice range in thickness from millimetres to several metres. Sediments consisting mainly of silt-sized particles (intermediate between clay and sand) are most susceptible to the formation of segregated ice. Coarser material lacks the necessary capillarity for water to be drawn towards the freezing pocket, whereas the low permeability of clay restricts water movement through the pore spaces. Where segregated ice forms it often takes on the shape of a lens lying parallel to the ground surface. As this **ice lens** grows, it pushes up overlying sediment causing doming of the ground.

## c) Frost cracking and ice wedge polygons

Once the ground is already frozen, further subzero temperature decline of the ground during winter causes contraction and cracking. This process of **frost cracking** creates fractures in the ground that extend from the surface through the active layer and into the permafrost. Seen from above, the cracks form an irregular polygonal pattern across the surface, with polygons typically 5–30 m across. The largest polygons are found where winter temperature falls below –20°C. The rate of temperature decline is thought to be more important for the intensity of frost cracking than the actual minimum temperature reached. During periods of thaw in the active layer, water seeps down and freezes into these cracks in the permafrost. As temperature declines during the following winter, contraction causes these same cracks to open up further so that in the next summer another layer of ice can form within a crack, adding to the previous layer. Over many years, this layering of ice within a crack produces an **ice wedge** that extends downward into the permafrost. Large ice wedges that are between 1 and 2 m wide at the top and between 8 and 10 m deep take around 100 years to form.

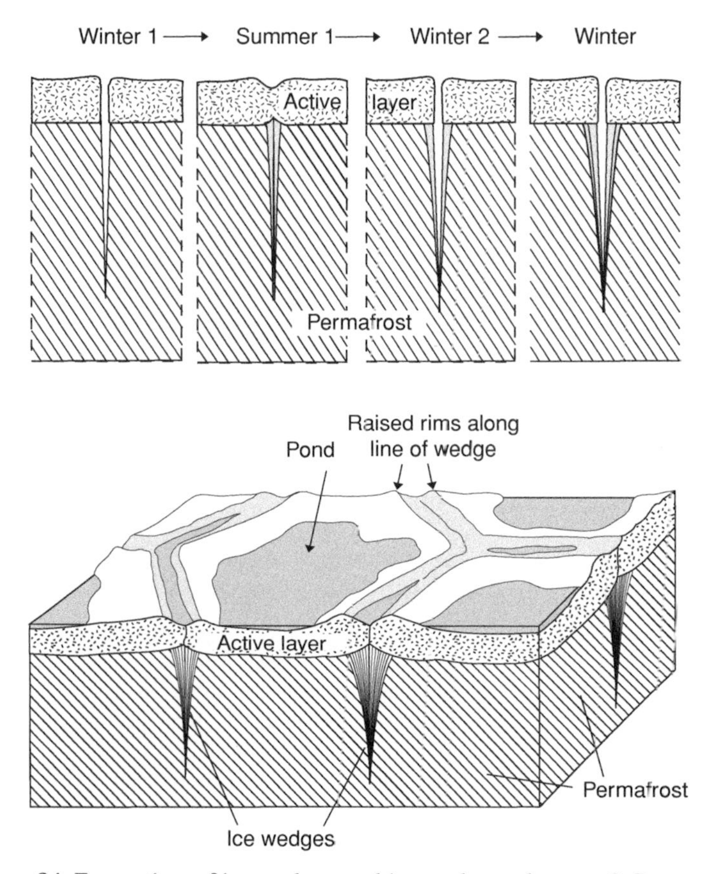


**Figure 24** Formation of ice wedges and ice wedge polygons. © *Process and Landform; An Outline of Contemporary Geomorphology 1st Edn.*, A Clowes and P Comfort. Reprinted by permission of Pearson Education Limited.

The polygonal cracking of the ground together with the formation of ice wedges within these cracks constitutes an **ice wedge polygon**. Active ice wedge polygons are found within the Arctic Circle, for example in the Mackenzie Delta and on Banks Island, Canada. Ice wedge polygons are bounded by raised rims of material running along both sides of the ice wedge (Figure 24). These rims are formed when expansion of sediment during autumn freezing causes material to be pushed upwards as it is forced against the more resistant ice wedge. The centre of an ice wedge polygon is lower, often containing a pond during the season of thaw in the active layer. If the ice wedges melt, they become filled with sediment to form **ice wedge casts**. Many ice wedge casts have been identified in areas of England, for example in East Anglia, that are beyond the limits of the last glaciation attesting to the former presence of permafrost.

## d) Patterned ground

**Patterned ground** is one of the most distinctive features of periglacial environments, and the ice wedge polygons described above are just one type of patterned ground among many others. Five main forms of patterned ground are recognised – circles, nets, polygons, steps and stripes. Patterned ground can be further classified by whether or not it exhibits particle size sorting. Although common in areas underlain by permafrost, patterned ground can also be found outside the permafrost zone.

Sorted patterned ground is characterised by the separation of stones (pebbles to boulders) from finer material on the surface so that the stones are organised into circles, nets, polygons, steps or stripes. **Frost heave** is a key process in the creation of sorted patterned ground because it provides a mechanism for separating stones from finer material. Broadly speaking, frost heave refers to the upward movement of sediment and soil due to expansion as water freezes within that material. Over many freeze–thaw cycles, frost heave causes stones to rise to the surface while finer material tends to move downward. This occurs for the following reasons.

- Stones have a lower specific heat capacity than finer sediments and therefore heat up and cool down more quickly. As the freezing front moves downward through the active layer in autumn it will move through a stone more rapidly than through surrounding material.
- This causes the top of the stone to become frozen to overlying material, which pulls the stone upwards with it as it expands (**frost pull**).
- As the freezing front will also penetrate beneath the stone relatively quickly, finer material underlying the stone will expand pushing the stone upward (**frost push**).
- During spring when thawing begins in the active layer, the stone stays in its uplifted position because fine sediments collapse into the cavity beneath the stone.
- Over successive years, these processes of frost pull and frost push cause stones to concentrate on the ground. A smaller scale version of frost heave can occur at or near the surface when freezing of the ground at night causes **needle ice** (or pipkrake) to form preferentially beneath pebbles lifting them off the ground by several millimetres or centimetres.

Once brought up to the ground, stones may become organised into sorted stone circles, nets, polygons, steps or stripes depending on the concentration of stones, the amount of moisture, the slope angle and the operation of slope processes. Stone circles, typically 0.5–3 m in diameter, can be found individually or in groups. While frost heave causes the stones to concentrate on the surface, **frost**

**thrust** causes the stones to move outwards from the centre to form the circle. In simplest terms, frost thrust occurs because the doming of the ground produced by frost heave causes stones brought to the surface to roll outward while fines are left in the central raised area. The interaction of many stone circles can lead to the formation of interlinked **stone nets** that can eventually grade into **stone polygons** reaching up to 10 m in diameter. Wider polygons tend to have larger stones along their borders than smaller polygons. These stone patterns are most symmetrical on flat surfaces, and they become elongated as the slope angle increases beyond about 2°. Steps and stripes occur at gradients between 5° and 30°. Beyond a gradient of 30°, mass movement is too active on the hill slope for the development of patterned ground.

**Steps** are bench-like features with downslope edges marked by an embankment of stones or vegetation, and they form where material pushed up from the ground by frost heave also moves downward by mass movement because of the slope gradient. Stone steps that resemble lobes of material oriented downslope are also referred to as **stone garlands**. With increased gradient, steps grade into sorted **stone stripes** that are characterised by lines of stones oriented downhill. The lines of stones, ranging in width between 0.3 and 1 m, are narrower than the lines of finer material in between. The relationship between slope angle and sorted patterned ground is illustrated in Figure 25.

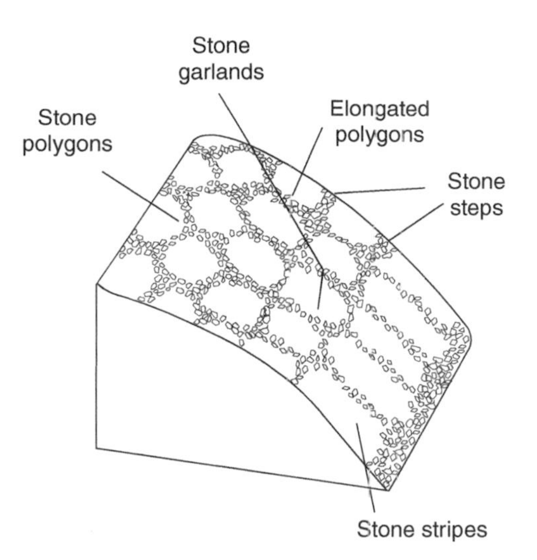


**Figure 25** The relationship between slope angle and sorted patterned ground

In addition to sorted patterned ground, there are many unsorted types such as the ice wedge polygons produced by frost cracking already discussed. The most common type of patterned ground is the non-sorted circle, also known as an **earth hummock**. These are simply circular mounds of vegetation that form an irregular pattern of hummocks over the surface, and they arise where frost heaving is concentrated into discrete areas beneath the ground. They can range from as little as a few centimetres in height to several metres.

## CASE STUDY: PATTERNED GROUND ON TINTO HILL, SCOTLAND

Sorted patterned ground is commonly found at altitudes exceeding 500–600 m in Scottish uplands where a maritime periglacial climate currently prevails. The abundant precipitation combined with temperatures fluctuating around 0°C in winter favours the processes of frost action and frost heave. This leads to the development of sorted circles, nets, polygons and stripes generally less than 1 m in width. Patterned ground of larger dimensions (up to 4 m width) is also sometimes found, but only as a relict feature formed during the Loch Lomond Stadial (12 800–11 500 years ago) when periglaciation was more intense and freezing penetrated deeper into the ground. Presently, frost sorting sorts stones that are normally less than 15 cm in diameter and produces smaller patterns than during the last glacial.

Tinto Hill in the Southern Uplands contains one of the best examples of active sorted patterned ground in Scotland (Figure 26). Tinto Hill is located 50 km south-west of Edinburgh and reaches an elevation of 711 m. To the south-east of the summit between 610 and 660 m, and at gradients between 15 and 25°, there are well-developed sorted stone stripes on unvegetated ground. The stripes are composed of alternating lines of coarse and finer debris running directly downslope. They range between 15 and 35 cm wide and between 10 and 30 m long, and they are made up of felsite debris derived from frost-shattering of bedrock around the hill's summit. Most of the stones in the coarse stripes range from 3 to 15 cm in diameter and have their long axes aligned downslope. The fine stripes contain black soil and gravels between 1 and 2 cm diameter.

The soil within the fine stripes has particle size and pore space characteristics that allow capillary movement of soil moisture toward centres of freezing. The soil is therefore said to be **frost susceptible**, and in winter, primarily, this causes ice lenses and needle ice to develop within the upper layers of the soil causing the entire fine stripe to become arched upward by several

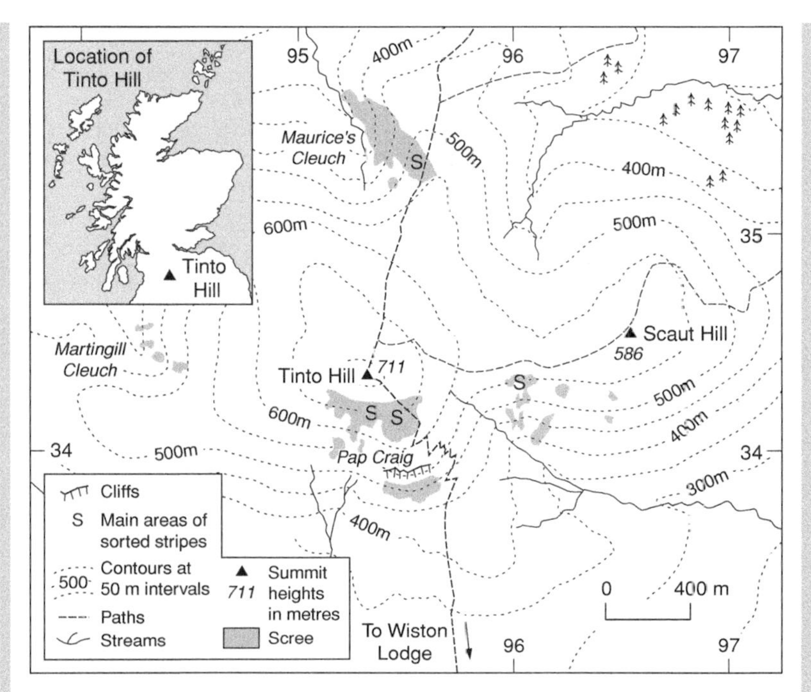


**Figure 26** Tinto Hill, Scotland

Sorted stone stripes south of the summit of Tinto Hill

centimetres. This process causes the stripes to have a ridge and furrow pattern, with fine stripes being ridged and coarse stripes forming furrows. This then maintains the process of lateral sorting (frost thrust), as larger stones brought up to the surface by frost heave will roll off the domed fine stripes to settle within the furrowed coarse stripes. While it is easy to see how the sorted stripe pattern maintains itself once established, it is more difficult to identify how the pattern is initiated. On Tinto Hill it appears that slight differences in soil texture and frost susceptibility near the top of the hill causes some patches of ground to experience more frost heave than other patches. At Tinto Hill this has created sorted nets with the most frost-susceptible soil near the centre and larger stones rolling out to the perimeter. With increased slope angle the nets elongate into stripes. The alignment of stones into stripes may also have been aided by slope wash and rill erosion during rainstorms. Observations of the stripes on Tinto Hill have shown that the larger stones move downslope by frost creep at rates of around 40 cm per year. These rates are far quicker than observed for stone stripes in most parts of the world and are due to the high soil moisture content and high frequency of frost heave that occurs under Scotland's maritime climate.

## e) Pingos

The term 'pingo' comes from an Eskimo word for hill. **Pingos** are isolated hills containing a core of ice, and they are roughly circular when viewed from above. They range in size from 30 to 600 m in diameter and from 3 to 70 m in height. Their summits are often rimmed with dilation cracks because of growth of the ice core and expansion of overlying sediment, but once the ice core is exposed at the surface it will begin to thaw in summer causing the top of the pingo to collapse inward, resembling a crater. Sometimes this collapsed area will contain a pond. Pingos develop slowly, with the largest pingos taking thousands of years to form.

There are two mechanisms to explain the formation of pingos, one involving a closed system process and the other an open system. **Closed system pingos** usually form from the isolation and progressive infill and disappearance of a small lake. Lakes have an insulating effect preventing the formation of permafrost in the underlying sediment. However, as a lake infills and shrinks, its insulating effect is reduced and permafrost will begin to encroach beneath the lake from the base and sides. The pressure of encroaching permafrost causes water in the sediment to concentrate beneath the former lake bed between the freezing fronts. This trapped body of groundwater eventually freezes, forming a mass of segregated ice that pushes up overlying sediment to create the pingo. Closed system pingos develop only in the zone of continuous permafrost. They are particularly well represented in the Mackenzie Delta area of the Canadian Arctic, where over a thousand have been identified. **Open system pingos** develop where a body of segregated ice is enlarged by groundwater flow. Instead of forming from trapped groundwater surrounded by permafrost, the open system model involves groundwater under artesian pressure moving through the permafrost to continually feed a centre of expanding ice that domes up overlying sediment. This type of pingo is often found in the discontinuous zone of permafrost where groundwater is able to circulate more easily through the subsurface sediment and rock.

### f) Thermokarst

**Thermokarst** is a general term referring to a landscape of topographic depressions due to the thawing of ground ice. It is characterised by extensive areas of irregular, hummocky ground interspersed with waterlogged hollows. The term came into use because the surface topography bears some resemblance to true 'karst' landscapes, which develop in limestone areas. Thermokarst depressions can take on a variety of dimensions depending on the extent and pattern of ground ice thaw. Depressions may fill with water to form **thaw lakes** that are relatively shallow, usually less than 5 m deep, and generally less than 2 km across. Narrow depressions of a polygonal pattern can occur because of the melting of ice wedge polygons. On the largest scale, **alases** are flat-floored and steep-sided thermokarst depressions that range in size from 3 to 40 m deep and from 100 m to 15 km long. They develop from the thawing of numerous ice wedges and other types of ground ice thereby causing large-scale ground subsidence, and they often contain lakes. The coalescence of individual alases results in **alas valleys** that can be tens of kilometres in length. The area around the Lena and Aldan rivers in central Yakutia, Russia, is a classic thermokarst landscape containing numerous alases and alas valleys.

The long-term thawing of ground ice that creates a thermokarst landscape can be initiated by climatic warming and/or by changes in the insulating characteristics of the surface. For instance, vegetation disturbance decreases the insulation of ground ice causing it to thaw to greater depths during spring and summer (extending the depth of the active layer). An increased rate of surface erosion also reduces insulation of underlying permafrost. Decreases in vegetation cover and increases in erosion that lead to thermokarst development can either be natural or due to human activity, as described in Chapter 7.

# 4 Periglacial Slope Processes and Related Landforms

Denudation is the wearing down of the Earth's surface by weathering and erosion, and in periglacial environments denudation is highly dependent upon frost action (freeze–thaw weathering) coupled with various types of mass movement occurring on hill slopes. Frost action is the dominant form of mechanical weathering because of the fluctuation of temperature above and below 0°C that is characteristic of periglacial climates. In cold climates chemical weathering is relatively unimportant, however in limestone areas the rate of carbonation weathering may be similar to that of warmer climates because of the increased solubility of carbon dioxide at lower temperatures.

## a) Blockfields, scree slopes and related landforms

The dominance of mechanical weathering under periglacial conditions produces regolith made up of angular rock fragments. An accumulation of frost-shattered stones and boulders on a relatively flat surface constitutes a **blockfield**. Blockfields are commonly found on mountain tops and plateaus unaffected by glaciation, but subject to periglaciation. They form *in situ*, produced by local freeze–thaw weathering and frost heaving of jointed bedrock. In some places rock that is more resistant to weathering may protrude above the surrounding blockfield as a **tor**. Where slopes are steeper, the frost shattered stones and boulders will move downhill forming block streams or stripes. These features are well represented in the upland granite landscape of Dartmoor, south-west England. This landscape is south of the limit of glaciation in Britain, and hence owes much of its present appearance to periglacial processes that operated during the Pleistocene glacials.

Along cliffs, rock fragments fall and accumulate at the base to form a **scree slope** (also known as a talus slope). The upper part of a scree slope typically rests at an angle between 30 and 38°, and the lower part rests at an angle between 25 and 30°. In general, a scree slope with a larger average particle size rests at a steeper angle than one made of smaller particles. However, the largest boulders tend to be at the base of the scree slope because greater mass and momentum allows them to travel further downslope. Some of the most impressive scree slopes in England are the Wastwater Screes found in the Lake District, which rise upward by more than 60 m from the shoreline of Wastwater Lake.

If there is a snow patch at the foot of a cliff, falling rock fragments will slide along the surface of the snow and accumulate at the base of the snow patch rather than directly beneath the cliff. Once the snow patch melts away, this leaves behind a rampart of boulders some distance in front of the cliff that marks the former edge of the snow patch and is called a **protalus rampart**. The Cwm Idwal cirque in Snowdonia (north Wales) contains a protalus rampart on its north side.

If a large quantity of frost-shattered rock becomes mixed with ice, a **rock glacier** is formed. This can occur either by a large supply of rock debris being added to a small and thin glacier or by the growth of ice within a large accumulation of rock fragments. The former type is referred to as an ice-cored rock glacier, whereas the latter type is known as an ice-cemented rock glacier. Rock glaciers tend to have a steep front, sometimes as much as 100 m in height, and they can reach a length of 1 km or more. The presence of ice between rock fragments allows the rock glacier to deform under its weight and move downslope at rates of up to 1 m per year.

## b) Frost creep, solifluction and related landforms

**Frost creep** and **solifluction** are two of the most important slope processes occurring in periglacial environments. Frost creep is a specific form of soil creep in which soil particles are pushed outwards at a 90° angle to the hill slope during freezing of interstitial water and then lowered downward along the vertical on thawing. Over many repeated cycles of freezing and thawing, this results in significant displacement of soil particles downslope. Frost creep is also effective at moving larger stones and boulders downslope. The process is slow, causing movement of at most a few centimetres per year on steeper slopes.

Unlike frost creep, which is a type of mass movement involving heave, solifluction is a mass movement characterised by flow. Solifluction refers to the downslope flow of water-saturated soil or regolith, and larger stones and boulders can also be carried along within this flow. Saturation causes a high pore water pressure between particles that reduces internal friction therefore enabling the whole mass of material to flow like a very viscous fluid. While faster than frost creep, it is still a slow process in which the rate of movement is normally between 0.5 and 5 cm per year, and rarely in excess of 10 cm per year.

Solifluction is a widespread phenomenon that is not restricted to periglacial environments. However, it commonly occurs under periglacial conditions because of the soil moisture provided by seasonal thaw. Where permafrost is present it creates an impermeable layer at a relatively shallow depth beneath the ground. In spring and summer when ice melts at the surface and within the active layer, the permafrost prevents downward percolation of meltwater causing it to saturate the upper layer of ground. This can cause solifluction on slopes as gentle as 1°, although maximum rates occur at gradients of 10–20°. At steeper angles more efficient drainage reduces the likelihood of soil saturation thereby limiting solifluction. The more specific term **gelifluction** is often used to refer to solifluction occurring in the permafrost zone.

The most distinctive landforms created by gelifluction include **gelifluction lobes** and **gelifluction benches**. The former refer to tongue-shaped deposits of gelifluncted material oriented downslope that tend to form on slopes of between 10 and 20°, whereas the latter refer to terrace-like deposits forming on gentler slopes with a long axis running parallel to the slope contour. Gelifluction lobes can be subdivided into either stone-banked or turf-banked lobes depending on the amount of vegetation cover. A lobe is measured according to the height of its front (the riser) and its length upslope (the tread).

Gelifluction lobes typically have risers of up to 5 m in height and treads 50 m long or more. Stone-banked lobes occur when there is a high concentration of stones at the surface because of frost heave. Examples can be found in many upland regions of Britain, for instance in Snowdonia on the mountains of Carnedd Llywelyn and Y Garn, and at Dartmoor on the south and south-east slopes surrounding Leather Tor. They represent the type of sorted patterned ground, also known as a stone garland, which occurs on moderate slopes intermediate between stone polygons and stone stripes. Turf-banked lobes have risers of soil and vegetation rather than stones because of a low concentration of stones in the gelifluncted material. Gelifluction benches are equivalent to the patterned ground referred to as steps. Like lobes, gelifluction benches can also be either stone-banked or turf-banked.

Gelifluncted material that forms an extensive low-angle deposit is known as a **gelifluction sheet**. In valleys experiencing periglacial conditions such sheets of material derived from surrounding slopes can build up to depths of several metres, and the gelifluncted deposit is referred to as **head deposit**. It usually contains a mix of fines, sand and frost-shattered stones that have long axes oriented downslope reflecting the flow direction. The chalk hills of southern England, including the North and South Downs and the Chilterns, were strongly affected by gelifluction during Pleistocene glacials because of the former existence of permafrost. In such areas the resultant head deposits contain a mixture of gelifluncted chalk and flint, and are known by the more specific term **coombe rock**.

Rates of frost creep and gelifluction can sometimes vary significantly on different sides of a valley leading to the development of an **asymmetric valley**. An asymmetric valley has one side markedly steeper than the other side. In periglacial environments asymmetric valleys are common, and their existence is often unrelated to the underlying rock structure (the orientation of bedding planes). In many cases their formation is related to slope aspect. A survey of slope angles in north-west Alaska showed that valleys trending approximately east–west tend to have steeper north-facing slopes. This can be explained for the following reasons:

- In the Northern Hemisphere, south-facing slopes receive more sunlight than north-facing slopes.

- This causes more thawing on south-facing slopes, which, by increasing soil moisture, promotes gelifluction and other forms of mass movement.
- Over time the removal of material by enhanced mass movement reduces the slope angle of the south-facing slope relative to the north-facing slope.
- The relative steepness of the north-facing slope may also be maintained because of stream undercutting as increased deposition at the base of the south-facing slope causes the meltwater stream to be diverted closer to the north-facing slope.

There are, however, other factors affecting weathering and mass movement in periglacial environments, and hence there are exceptions to a steeper north-facing slope. Patterns of snow accumulation and vegetation cover also play an important role, as does the regional and local climate regime. For example, in extremely cold, high-latitude periglacial environments the south-facing slope can become the steeper slope because the additional solar energy allows for temperature fluctuation around 0°C enabling freeze–thaw weathering, rock fall and the development of a free face.

### c) Nivation and related landforms

**Nivation** is a process of localised weathering and erosion beneath and around a snow patch that creates and enlarges a hollow on a hillside. The presence of a snow patch provides abundant moisture accelerating frost action, frost creep and gelifluction. Furthermore, meltwater from the snow patch in spring and summer helps to transport weathered material away from the developing hollow. In limestone areas, carbonation weathering may also be accelerated by the presence of the snow patch. Nivation was introduced in Chapter 3 because it is the first stage in the formation of a glacial cirque. However, snow patches do not always develop into glaciers, and in the absence of glaciation the **nivation hollow** (or **nivation cirque**) remains considerably smaller than a glacial cirque.

In addition to forming hollows, nivation can lead to cliff recession where many snow patches persist along the base of a cliff. Nivation weathers and undermines the base of the cliff, and gelifluction and slope wash transport weathered debris away. Over a long period of time this can produce a gently sloping rock terrace extending away from the cliff known as a **nivation terrace** (or **cryoplanation terrace**).

## 5 Action of Water and Wind

Owing to the open and sparsely vegetated landscape characteristic of periglacial environments, rates of erosion caused by water and wind can be high. Water erosion is highly seasonal, occurring mainly in spring and summer when snow and ice are on the surface and in the active layer melts. This can cause short periods of very high discharge, causing far more fluvial erosion than would be expected given the relatively low mean annual values of precipitation and discharge characteristic of periglacial environments. As described in Chapter 5, near the margins of glaciers drainage is typically braided because of the high amount of debris being carried by meltwater streams. Where streams are not so highly loaded with sediment, the drainage pattern is more like that of temperate climates, although active for much shorter periods of the year.

The presence of permafrost, or seasonally frozen ground, prevents percolation of water downward into the bedrock forcing it to flow over the surface or at shallow depths in the ground. This results in a relatively high drainage density that can produce a network of valleys. In areas underlain by permeable rock, particularly chalk and limestone, **dry valleys** provide important evidence for the former existence of ground ice. The dry valley can therefore be classed as a relict periglacial landform. The chalk hills of southern England provide particularly good examples of this feature. For instance, near the village of Brook, close to the chalk escarpment of the North Downs, there is a cluster of seven such dry valleys (also known as coombes). When the landscape of southern England consisted of tundra and ground ice during the last glacial and earlier glacials, meltwater produced during the summer thaw was unable to percolate down into the chalk. Instead, gelifluction and meltwater erosion created valleys in the chalk. With climatic warming at the beginning of the present interglacial, permafrost disappeared from England, and water was again able to percolate into the rock leaving the dry valleys seen today.

Unobstructed winds blowing across periglacial landscapes can reach high velocities causing erosion through **wind abrasion** and by dislodging fine, unconsolidated materials through the process of **deflation**. The effects of wind abrasion can be seen in grooved and polished rock surfaces and in stones shaped by the wind known as **ventifacts**. During Pleistocene glaciations, a great deal of fine material (mainly silt-sized) from outwash plains along the margins of continental ice sheets was picked up by the wind and carried long distances to be deposited elsewhere as extensive areas of **loess**. Loess is a fine-grained, unconsolidated wind-blown deposit, and it can cover large areas to depths of several metres. Loess is found in many parts of North America and Eurasia south of the boundary of Pleistocene ice sheets. In England loess deposits cover parts of East Anglia and the London Basin where they are referred to as brick-earth deposits, although they rarely exceed 2 m in depth. Loess deposits become deeper and more extensive further east because of the influence of prevailing westerly winds. China contains the deepest loess deposits, which exceed 300 m in places. The soils derived from loess deposits are excellent for arable agriculture because of their loamy texture and high fertility. Coarser periglacial wind deposits consisting mainly of sand-sized particles are known as **coversand**.

## Summary Diagram

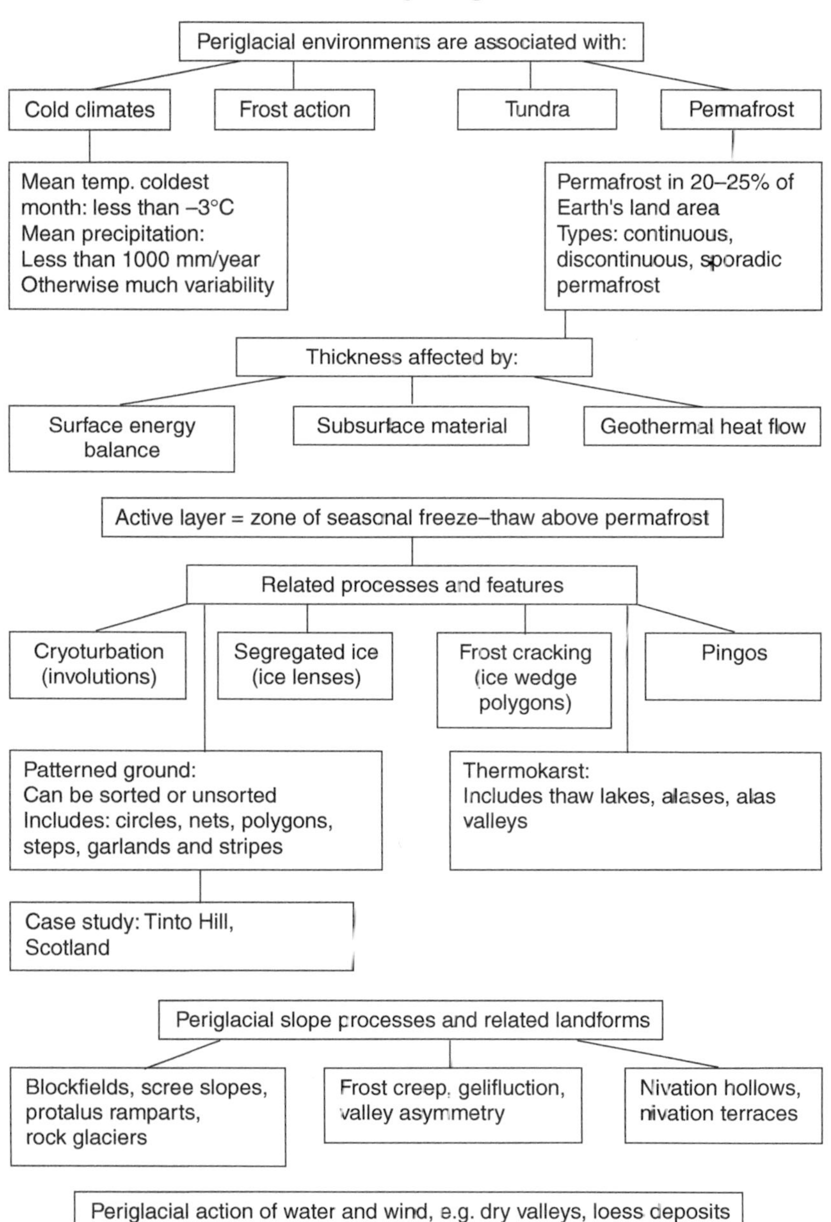

## Questions

1. **a)** Describe the distribution of periglacial environments.
   **b)** The distribution of periglacial environments and the distribution of permafrost are similar but not identical. Why not?
2. **a)** Describe the processes that operate in the active layer.
   **b)** Explain why stones tend to move upward through the active layer.
   **c)** What factors contribute to the development of sorted patterned ground?
3. Describe and explain the periglacial features and landforms associated with ground ice.
4. **a)** Describe the slope processes that occur in periglacial environments.
   **b)** What distinctive features result from these processes?

# 7 Human Activity and Glacial and Periglacial Environments

**KEY WORDS**

**Anthropogenic climate change**: climate change caused by human activity, particularly referring to greenhouse warming due to increased emissions of greenhouse gases.
***Jökulhlaup***: an Icelandic term for a glacial outburst flood involving the sudden discharge of a subglacial or ice-dammed lake.
**Transhumance**: a type of pastoral agriculture in which livestock are grazed in the highlands in summer and moved to lower ground in winter.
**Utilidor**: an insulated box conduit elevated above the ground that protects water supply, heating and sewage pipes from frost damage and prevents permafrost degradation.

## 1 Opportunities and Constraints of Glacial Environments

At first glance it might seem that glacial environments present nothing but difficulties for human activity. While the climate and landscape are indeed harsh, there can, nonetheless, be many benefits gained from adapting to and utilising such environments. Before outlining various opportunities and constraints for human activity, it is important to make the distinction between landscapes that are currently being glaciated and those that were glaciated in the past. Strictly speaking, a glacial environment should be thought of as an area that is currently experiencing glaciation. However, in the discussion that follows, opportunities and constraints will also be considered for formerly glaciated areas because the effects of past glaciation on human activity can also be profound.

### a) Tourism

The tourism industry has seen tremendous growth in recent decades, and this has brought many economic benefits to mountain regions. Visitors are attracted by the spectacular scenery of jagged peaks and deep valleys produced by glacial erosion. Pyramidal peaks, arêtes, cirques and hanging valleys are just some of the more impressive features of such landscapes. The scenic value of upland regions in Britain, such as the Lake District and Snowdonia, is the result of past glaciation, whereas in many other mountainous regions, such as the

Alps, glaciers are still shaping the land at high altitudes. The rugged landscape of glaciated areas provides opportunities for hill walking, climbing, mountaineering and skiing, and many towns and regions have capitalised on the growing popularity of these activities. For example, Aspen, Colorado began as a silver mining settlement in the 19th century but since the 1960s it has become one of the pre-eminent ski resorts in the world.

In addition to outdoor recreation, glaciated regions are increasingly being visited for the glaciers themselves. In the Bernese Oberland of Switzerland, Jungfrau Railways transports tourists up to the Jungfraujoch, which, at 3454 m altitude, is the highest railway station in Europe, offering spectacular views across the Aletsch Glacier. Near Chamonix in the French Alps tourists can visit the Mer de Glace Glacier and enter an ice cave within the glacier. Also near Chamonix, a cablecar up to the Aiguille du Midi observation deck at 3842 m offers panoramic views of Mont Blanc and surrounding glaciers. In New Zealand, visitors can join guided tours of the Franz Josef Glacier in Westland National Park. In Canada, the Columbia Icefield Centre next to the Athabasca Glacier, and the Brewster Snocoach tour onto the glacier, provide yet further examples of tourist provision for learning about and viewing glaciers.

Increased levels of tourism in glaciated areas also present problems. Mountain ecosystems are easily damaged by human impact, and excessive development on steep slopes can create instability, thereby increasing the hazards of mass movement. Conflicts of interest can occur where tourist developments compete for space with more traditional mountain land uses, such as pastoral agriculture and forestry. The scenic value and quality of experience for the individual tourist can also be undermined by overdevelopment and overcrowding. The dangers of mismanaging mountain environments were recognised in the United Nations Conference on Environment and Development (UNCED) in 1992 with the establishment of the Mountain Agenda. This programme has sought to gather data, to educate and to bring together different interest groups for the purpose of achieving more sustainable development in mountain regions.

## b) Water supplies and energy

Glaciated uplands receive relatively high amounts of precipitation compared with lowlands, and glaciers supply meltwater, which often collects in tarns or ribbon lakes. Glacially eroded valleys in many mountain regions have therefore been dammed and turned into reservoirs to control the supply of water to larger settlements located at lower altitude and to generate hydroelectric power (HEP). This is well exemplified in Switzerland where about 60% of the country's electricity is from HEP. The country contains over 500 hydroelectric power stations as well as the third tallest dam in the world, the Grande

Dixence dam in canton Valais, which was built across a glacial trough in 1961. This dam is 285 m high and 695 m long, and it impounds a reservoir behind it (Lac des Dix) that contains about 400 million $m^3$ of water.

## c) Agriculture

Agriculture in upland glaciated regions is primarily pastoral because the high precipitation, steep terrain and shallow soils are unsuitable for arable cultivation. The traditional pattern of livestock rearing in the Alps is termed **transhumance**, and this farming system takes advantage of the seasonal vegetation cycle in mountain environments. In summer animals are grazed at higher altitudes when the alpine meadows are free of snow and at their most productive. During this time pastures in the valley can be used to make hay and silage for animal feed in the winter. In autumn and spring animals are grazed at lower altitudes or in the valley bottom pastures.

In formerly glaciated uplands of Britain, such as the Lake District, the pattern of sheep farming is similar to that described for cattle in the Alps. The open fell is the highest altitude zone beginning above 300 m, which is used for grazing in summer. Lower down is the intake, which is a zone of enclosed and improved land that is used when conditions are poor up on the open fell, and along the valley floor is the most fertile and productive zone known as the inbye. The inbye is used for making hay and silage, and for grazing during winter or when the sheep need to be nearby for lambing and shearing. Where glaciated uplands are not used for sheep farming, they are often used for **silviculture** (forestry). Large areas of the Scottish Highlands, for example, are planted with non-native, quick-growing conifer species such as Lodge pole pine and Sitka spruce for timber and wood pulp. These conifers are able to tolerate harsh climatic conditions and thin, acidic soils that would not be suitable for other types of land use.

In contrast with upland areas, lowland glaciated regions in the mid-latitudes can be well-suited to arable agriculture. Where Pleistocene ice sheets have left behind extensive till plains, the ground surface is relatively flat and therefore energy efficient for driving tractors and other farm machinery. The till (or boulder clay) can form the basis for a rich agricultural soil with good water retention. However, because of the high clay content of soils based on glacial till, these soils sometimes retain water too well and therefore require artificial drainage to prevent waterlogging. In Britain, East Anglia provides the best example of an area of glacial till that has been successfully used for arable agriculture. Deposits of lodgement till from successive glaciations have built up into a thick till plain that covers much of East Anglia. Chalk from the underlying bedrock mixed in with the overlying till forms the characteristic 'chalky boulder clay' soil of the region.

Combined with a relatively low mean annual precipitation of around 600 mm, the chalky boulder clay soil is highly suitable for cereal cultivation.

## d) Mining and quarrying

Mountainous areas offer opportunities for mining because the tectonic forces creating the mountains result in folds and faults that expose various rock layers and mineral seams at the surface. Glacial erosion plays the additional role of removing overlying vegetation and regolith to more clearly expose economically valuable rocks and minerals at the surface once glaciers have retreated. Over long periods of time successive glaciations remove enough material to promote **dilatation** of bedrock (pressure release weathering) that opens up joints and fractures facilitating the access and removal of rocks and minerals. For instance, the glaciated uplands of North Wales once had a thriving slate mining industry. At its height in the 1890s this industry employed nearly 17 000 people and exported Welsh slate all over the world. The evidence for large-scale slate mining in the region is amply demonstrated by the numerous slate waste heaps found along many of the mountain slopes. During the 20th century the industry declined dramatically, although it still continues at a smaller scale today.

In lowland areas, outwash deposits from Pleistocene ice sheets provide an important source of sand and gravel for the construction industry. It is the sorted character of outwash deposits that makes them economical to quarry because the sand and gravel can be extracted without needing time-consuming separation from smaller and larger particle sizes. These sands and gravels are then sold as aggregate for use in making concrete, mortar and infill for a wide variety of construction projects ranging from road building to housing developments.

## e) Settlement

Glaciated areas present both opportunities and constraints for many kinds of settlement. Of course, upland glaciated areas pose difficulties for settlement because of the rugged terrain and lack of flat space for building. Infrastructure for transport and communications is also more costly to build and maintain, although less so where glacial diffluence has created mountain passes. However, these constraints can also present certain kinds of opportunities. Historically, glaciated terrain has provided opportunities for defence and/or for controlling the movement of people and goods through an area. More recently, settlements in upland glaciated areas have expanded to capitalise on new economic advantages, such as the rise in tourism already discussed.

In lowland areas the distribution of glaciofluvial deposits, forming river terraces, can be an important factor in the location of settlements. Within river floodplains Pleistocene river terraces provide areas of raised and relatively flat ground where settlements can develop free of regular flooding, while also having good access to a river. This certainly applies to many settlements along the River Thames. Central and north Oxford is built on a river terrace that stands several metres above the present floodplain. The terrace made a convenient river crossing and dry-point site, which increased in importance during Saxon times when Oxford was located on a major trade route between the kingdoms of Mercia and Wessex. This terrace is part of the Summertown–Radley Terrace system, which is also found in many other places within the Upper Thames valley. It is composed of limestone, quartzite and flint gravels, and most of it is believed to have been deposited when England was experiencing glacial and periglacial conditions.

## 2 Hazards in Glacial Environments

Glaciated and previously glaciated upland regions are hazardous because of the high incidence of avalanches, rock falls, and other forms of mass movement and flooding. A natural hazard is a naturally occurring event that presents a risk to life or property. A disaster can be thought of as the realisation of this risk, and the scale of a disaster depends greatly on the degree of human vulnerability as well as on the magnitude of the natural event. Human vulnerability has been rising in glacial environments because of increased population and development as well as the growing popularity of skiing and other mountain sports. As a result, more people are at risk from these hazards than ever before.

### a) Avalanches

Avalanches are discussed in detail in a companion volume *Hazards*. While the vast majority of avalanches are too remote to pose any risk to humans, there are mountainous areas where vulnerability has risen, largely because of increased tourist development. On average around 200 people are killed in avalanches every year, with over half of these deaths in the Alps.

### b) Glacial outburst floods

A glacial outburst flood is also known by the Icelandic term ***jökulhlaup***. It is a powerful flood caused by the sudden discharge of a subglacial or ice-dammed lake. The catastrophic flooding from glacial Lake Missoula described in Chapter 5 provides an example of a glacial outburst flood on the largest scale. There is potential for an outburst flood wherever meltwater collects behind an ice obstruction, and sudden release of water can be triggered in a variety of ways. These include:

- increased flotation of ice as water levels rise
- overflow and melting of an ice dam
- break down of an ice dam because of tectonic activity
- enlargement of pre-existing tunnels beneath an ice dam because of increased water pressure.

Outburst floods can also be cyclic in nature. Following a flood, the ice dam may reform and the ice-dammed lake will gradually fill over several years until it reaches the critical level necessary to burst through the obstruction again.

Glacial outburst floods present a hazard in many glacial environments where settlements or structures are located down-valley of glaciers. They are particularly frequent in Iceland because of volcanic activity that both generates meltwater beneath glaciers and acts as a trigger for ice instability and subsequent release of meltwater. The Vatnajökull ice cap located in south-east Iceland covers about 8% of the country and averages 400 m thick, and it is the source of large outburst floods that occur fairly regularly over time. Heat from the Grimsvötn volcano beneath the ice cap melts ice and creates a subglacial lake within its caldera. When the subglacial lake reaches a critical size it forces its way through subglacial tunnels and the lake drains catastrophically in a matter of hours. This process produces outbursts on average every 5–6 years, with total discharges usually ranging between 0.5 and 3.5 km$^3$. In 1922 an outburst flood occurred that discharged approximately 7 km$^3$ of water and, for a short time, had a maximum flow about a quarter of the flow of the Amazon River. More recently, on 30 September 1996, an eruption of Grimsvötn broke through the ice sending up a column of ash 10 km high and melting a large quantity of ice. The meltwater remained obstructed by ice until 5 November when it burst through the glacier. This was the largest flood from Grimsvötn since 1938 with a massive peak discharge of 55 000 m$^3$ s$^{-1}$, and it severely damaged bridges, roads and power lines causing an estimated US$15 million of economic loss.

Besides Iceland, glacial outburst floods pose a high risk in several other parts of the world. In Washington State, USA there is a high potential for outburst floods from Mount Rainier, and this presents a serious threat to the area around Seattle. Mount Rainier has 25 glaciers radiating from its summit, containing more ice than any other mountain in the USA outside Alaska. Outburst floods from the mountain have occurred many times in the past century because of periods of relatively high temperature and rainfall that generated excess meltwater. Human-induced climatic change (global warming) could cause more and, perhaps, larger outburst floods of this type in the future. Mount Rainier is also a volcano, and an eruption would melt far more snow and ice causing immense flooding downstream. By mixing with sediment and soil, the flood waters from an eruption would produce a **lahar** (mudflow) capable of travelling a distance over 50 miles from the volcano. Studies of lahar deposits around Mount Rainier have shown that past lahars have extended well into areas near Puget Sound that are now heavily populated.

There is also growing concern about glacial outburst floods in the Himalayas. In 1985 the Dig Tsho glacial lake in Nepal burst through its confines destroying 14 bridges and causing about US$1.5 million of damage to a hydroelectric power plant downstream. However, this disaster was relatively small scale compared with predictions for future glacial outburst floods in the region. Recent climatic warming is believed to have significantly increased melting and retreat of Himalayan glaciers, and many glacial lakes in Nepal and Bhutan have become dangerously full. One of these lakes, Tsho Rolpa lake in the Dolakha district of Nepal, has grown to six times its size in the late 1950s. An outburst from this lake would damage a large amount of infrastructure and agricultural land downstream and would threaten the lives of over 10 000 people. Engineers are developing and installing sensors and sirens in the hazardous zone in the hope that in the event of an outburst flood people will have early enough warning to evacuate. Scientists from the United Nations Environment Programme (UNEP) have estimated that more than 40 other lakes in Nepal and Bhutan have become similarly unstable and could experience outbursts in the next several years.

Along the Andes in the tropics, muddy floods and ice/rock avalanches from high-altitude glaciers are sometimes referred to as **alluviones**. Similar to *jökulhlaups*, they can be caused by sudden release of meltwater from a formerly blocked glacial lake. They can also be caused by earthquakes that trigger avalanching of ice and rock. The Cordillera Blanca mountains of Peru have been the source of some devastating alluviones. In 1970, for example, an earthquake-induced ice and rock avalanche from Mount Huascaran travelled 11 miles from its source, burying the towns of Yungay and Ranrahirca, and killing about 18 000 people.

## 3 Challenges and Problems of Periglacial Environments

Although periglacial climates vary in their degree of severity, conditions are generally harsh, undoubtedly presenting a challenge for human occupation and development. Through history, the peoples inhabiting periglacial environments have adapted their cultures in remarkable ways to cope with the climate and to make the most of the limited resources. For instance, the traditional way of life of the Inuit (Eskimos) of Canada and Greenland was semi-nomadic and subsistence based, closely adapted to the environment and mainly dependent upon hunting and fishing. From the sea, whales, walrus and seals were important sources of food, skins and fuel; and on land they hunted many types of mammals including moose, caribou, musk ox and bear.

Establishing permanent settlements and developing industries, such as mining and oil extraction, has required major technical advances to be successful primarily because the presence of permafrost creates a unique set of problems for construction work and engineering. Permafrost underlies about 80% of Alaska and large tracts of northern land in Canada, Russia and Scandinavia, and the economic development of these areas has depended upon specially adapted buildings and infrastructure. Conventional construction techniques are unsuitable because they alter the thermal balance of the ground leading to permafrost thaw and ground subsidence. Problems are also caused when vegetation is cleared from the surface. This reduces the insulation of the permafrost, and in the summer this results in deepening of the active layer as heat is transferred down to the permafrost table more easily.

Ground subsidence is most severe in places where the permafrost is ice-rich, particularly in fine-grained sediments with a high porosity, so that deepening of the active layer causes a large volume of ground ice to melt relative to the original frozen volume of sediment. Even minor vegetation disturbances, for example by off-road vehicle tracks, can greatly increase melting of ground ice over the long term because tundra vegetation is very slow to re-establish itself. The effect of reduced insulation is exacerbated by structures, such as buildings or pipelines, that transmit additional heat to the ground. These effects speed up the development of a thermokarst landscape beyond the natural rate.

The damage caused by this form of ground subsidence can be seen in towns and villages built in permafrost areas before special engineering designs were developed. Historic buildings in the town of Dawson in the Yukon Territory of Canada, hub of the Klondike gold-rush of the late 19th century, are tilted and fractured. Roads, railways and airstrips in many parts of the Arctic have also been damaged by ground subsidence. In Canada, Sachs Harbour airstrip on southern Banks Island has required constant maintenance and re-levelling since it was completed in 1962 because the underlying sediments are ice-rich and the area has experienced rapid development of thermokarst. Bridges are often damaged by frost heave because middle supporting piles are driven into unfrozen sediments beneath stream beds, whereas piles on either bank are driven into the active layer. Consequently, piles at the ends of the bridge are pushed upward due to seasonal frost heave, while piles in the middle remain stationary because of the absence of frost heave beneath the body of water.

In recent years many new methods of construction have been employed to protect permafrost and to prevent subsidence. While largely successful, these methods are initially more expensive than conventional construction, adding to the cost of living in permafrost areas, and continual maintenance of the structures is often necessary. Houses and other small buildings are now often elevated above the ground on piles driven into the permafrost. A gap about 1 m high beneath the base of the building and the ground surface allows air to circulate and remove heat that would otherwise be conducted into the ground. Larger structures are built on **aggregate pads**, which are layers of coarse sand and gravel, typically 1–2 m thick, that substitute for the insulating effect of vegetation and also reduce transfer of heat from the building into the ground. Aggregate pads are also used beneath roads, railway tracks and airstrips. The thickness of the pads must be carefully calculated to maintain the thermal balance of the ground at its natural state. If too thin, insulation may not be strong enough to prevent thaw of permafrost causing subsidence, whereas if too thick, increased insulation may result in elevation of the permafrost table and upward heaving of the ground.

Another challenge in permafrost areas is the provision of water and energy supplies and waste disposal. Telephone and electricity cables and water pipes cannot be buried underground because of the stresses and damage that would be caused by freezing and thawing in the active layer. Telephone and transmission poles require continual maintenance because of frost heave. In relatively large settlements of a few thousand people or more (such as Inuvik in northern Canada)

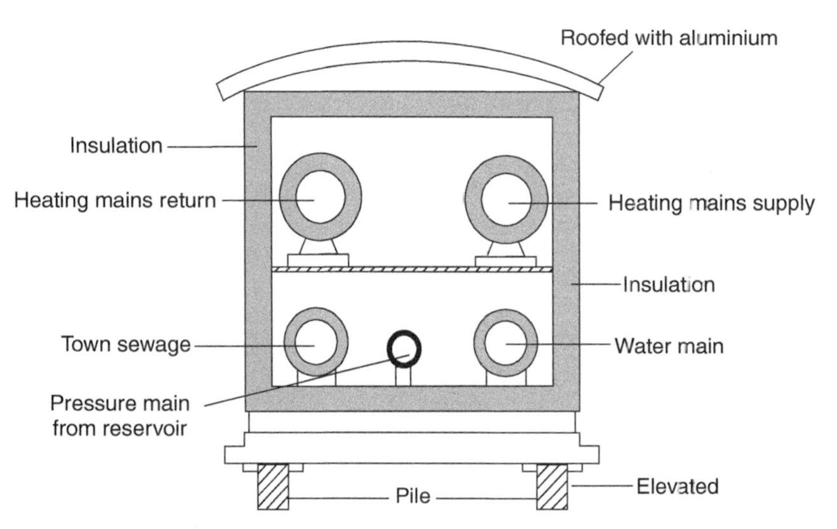


**Figure 27** Cross-section of a utilidor. *Source*: Money (1980).

**utilidors** are used to carry the water supply, heating pipes and sewers between buildings. Utilidors are insulated box conduits made of concrete, wood or metal that are elevated above the ground (Figure 27). In smaller settlements where it is uneconomical to set up a utilidor system, water supply and waste disposal are provided by truck.

## CASE STUDY: THE TRANS ALASKA PIPELINE

Since the 1970s increasing effort has been made to exploit oil and natural gas in Arctic regions of Alaska, Canada and Russia. This too has its unique challenges relating to exploration, drilling and transport in a periglacial environment. For example, oil extracted from the Prudhoe Bay oilfields in Alaska must be transported 1290 km (800 miles) to the ice-free port of Valdez before it can be loaded onto tankers and shipped to market. This is achieved by the Trans Alaska Pipeline, which was completed in 1977 and is a remarkable feat of engineering. It is very important for the United States oil industry, delivering 17% of its domestic oil production. It took approximately 6 years to design and about 3 years to build, employing about 70 000 people over the duration of construction and costing around US$8 billion in total. This was far in excess of the original cost estimate, largely because of unforeseen difficulties in building the structure across extensive areas of permafrost. The pipeline has a maximum daily throughput of some 1.4 million barrels, and the oil flowing through the pipeline is at a temperature of 65°C.

The Trans Alaska Pipeline has a number of design features that are specially adapted to the permafrost that is encountered along about 75% of its length. Where the pipeline must cross areas of fine-grained, ice-rich permafrost sediment (just over half its length) it is elevated above the ground so that heat from the oil is not conducted into the ground. This is important to prevent thaw of ground-ice because this would cause subsidence and solifluction of soil, thereby damaging the pipeline. If the pipeline were to rupture it would cause enormous ecological damage. The elevated structure is highly sophisticated, allowing the pipeline to shift sideways on its supports as an extra protection from ground movement, and the pilings are specially designed to resist being 'jacked up' by successive years of frost heave. The elevated pipeline is also built in a zigzag pattern, rather than in a straight line, so that it is able to adjust to ground movements caused by temperature changes or even by earthquakes. Where the pipeline must be buried because of roads, animal crossings and avalanche hazards, thick insulation coatings are used, and in the most sensitive permafrost areas

The Trans Alaska Pipeline

refrigeration pipes are installed around the main pipeline to keep the ground frozen.

In areas free of permafrost, or where the permafrost sediment is coarse grained (free draining and less susceptible to subsidence on thawing), the pipeline is buried underground using cheaper, conventional methods. Designing the pipeline in the most cost-effective way demanded careful study of the different soil and sediment types along its length, and their degree of susceptibility to ground subsidence and frost heave. Only about 6.5 km of buried pipeline required the most expensive design including below-ground refrigeration.

# 4 Human Impact on Glacial and Periglacial Environments

There are two main ways by which humans have an impact on glacial and periglacial environments – pollution and development, and **anthropogenic climate change**.

## a) Pollution and development

The low temperatures of glacial and periglacial environments slow processes of chemical degradation and organic decomposition making these environments particularly vulnerable to pollution. For instance, the ecological damage caused by oil spills is much longer lasting in the polar latitudes than elsewhere because oil degrades much more slowly under cold conditions. Most of the other pollutants found in polar regions are from distant sources; however, they can easily accumulate within the food chain. Various industrial and agricultural chemicals, heavy metals and radioactive particles from lower latitudes have been found in Arctic ecosystems where, through **biomagnification**, they become concentrated in long-lived animals, such as seals, whales and polar bears, that feed near the top of the food chain. Polychlorinated biphenyls (PCBs) used in paints, plastics and electrical equipment are a common pollutant in marine ecosystems in the Arctic, while agricultural pesticides, such as HCH and DDT, are widely detected in the terrestrial ecosystems. These pollutants can have severe effects on wildlife, particularly by impairing reproduction. Acid precipitation (or 'acid rain') from burning fossil fuels can also be a problem in periglacial environments. Although rates of sulphate deposition are usually low in areas far removed from industrial activity, acid precipitation can build up over time and be stored in the snow pack to be released suddenly during thawing in spring. This can cause rapid acidification of lakes, particularly in areas where the soil and bedrock have a low acid-buffering capacity.

As already described, development schemes in periglacial environments can cause degradation of permafrost as heat from buildings and industrial activity is conducted into the ground. This can be thought of as thermal pollution and, if unchecked, over the long term it produces boggy, thermokarst-type terrain that is difficult to build upon and has little economic value. Tundra vegetation is also damaged easily by off-road vehicles, and once the vegetation cover is disturbed or removed it takes a long time for it to re-establish because of the short growing season and low ecological productivity.

## b) Anthropogenic climate change

Throughout the Pleistocene, glacial and periglacial environments have undergone repeated phases of expansion and retreat as the

Earth has shifted between glacial and interglacial climate conditions (Chapter 1). The natural causes for these glacial/interglacial cycles, particularly changes in the path of the Earth's orbit, continue to operate, and therefore over the long term it is likely that there will be many more such oscillations. As in the past, glacial and periglacial environments will continue to change in response to natural climatic variability. However, the extent to which anthropogenic climate change will modify these natural climatic trends in the short term is the focus of intense study and debate. Humans have been evolving throughout the Quaternary Ice Age, although it is only in the last few hundred years of the present interglacial that we have gained the capacity to significantly alter the global climate via the effects of industrial and agricultural activity on the atmosphere.

The human-caused increase in the atmospheric concentration of carbon dioxide, methane and other greenhouse gases is thought to be partly responsible for the 0.6°C rise in the Earth's average surface temperature over the 20th century. As described in the case study of Chapter 2, this has led to glacier retreat in most of the world's mountain regions. The average global snow cover has decreased by about 10% since the 1960s, and the extent of Northern Hemisphere spring and summer sea-ice has decreased by 10–15% since the 1950s. Remarkably, in recent decades there has been about a 40% decrease in Arctic sea-ice thickness during late summer.

In the 21st century these trends will continue as human activities add more greenhouse gases to the atmosphere. The Intergovernmental Panel on Climate Change (IPCC) estimates that by 2100 AD average global surface temperature will be between 1.4 and 5.8°C higher than in 1990. Even taking the most conservative estimate, alpine glaciers are likely to continue their rapid retreat. Land areas, particularly during winter at northern high latitudes, are predicted to warm more than the global average. Over the next century it is also predicted that the Greenland ice sheet will lose mass, although the Antarctic ice sheet will probably gain mass because under a warmer global climate Antarctica will receive enough extra precipitation to offset increased ablation. Average sea level could rise almost a metre by 2100 because of thermal expansion of the oceans and net ablation of the world's glaciers.

In summary, anthropogenic climate change is at least partly responsible for the current rate of retreat of alpine glaciers. If the world continues to warm in the 21st century as projected, the zones of periglaciation and permafrost are also expected to retreat to higher altitudes and latitudes. An intriguing question is whether the greenhouse warming will be of sufficient magnitude and duration to prolong our present interglacial, which has so far lasted for 11 500 years. Evidence from ocean sediments (Chapter 1) suggests that the average length of an interglacial is about 11 000 years, and therefore our warm interglacial should be nearing its end. Over the short term it would seem that greenhouse warming may prevent this. However, there is also the possibility that it could eventually trigger cooling, perhaps even hastening the onset of the next glacial, through disruption of the current system of ocean circulation that transports heat toward the poles. This could be caused by excess freshwater input into the North Atlantic Ocean (from increased precipitation and snow/ice melt at high latitudes) interfering with the northward flowing warm currents, in a manner similar to that described for the Heinrich events discussed in Chapter 2. More needs to be learnt about the complex interactions within the climate system before we can predict exactly how anthropogenic climate change will affect glacial and periglacial environments in the future.

## Summary Diagram

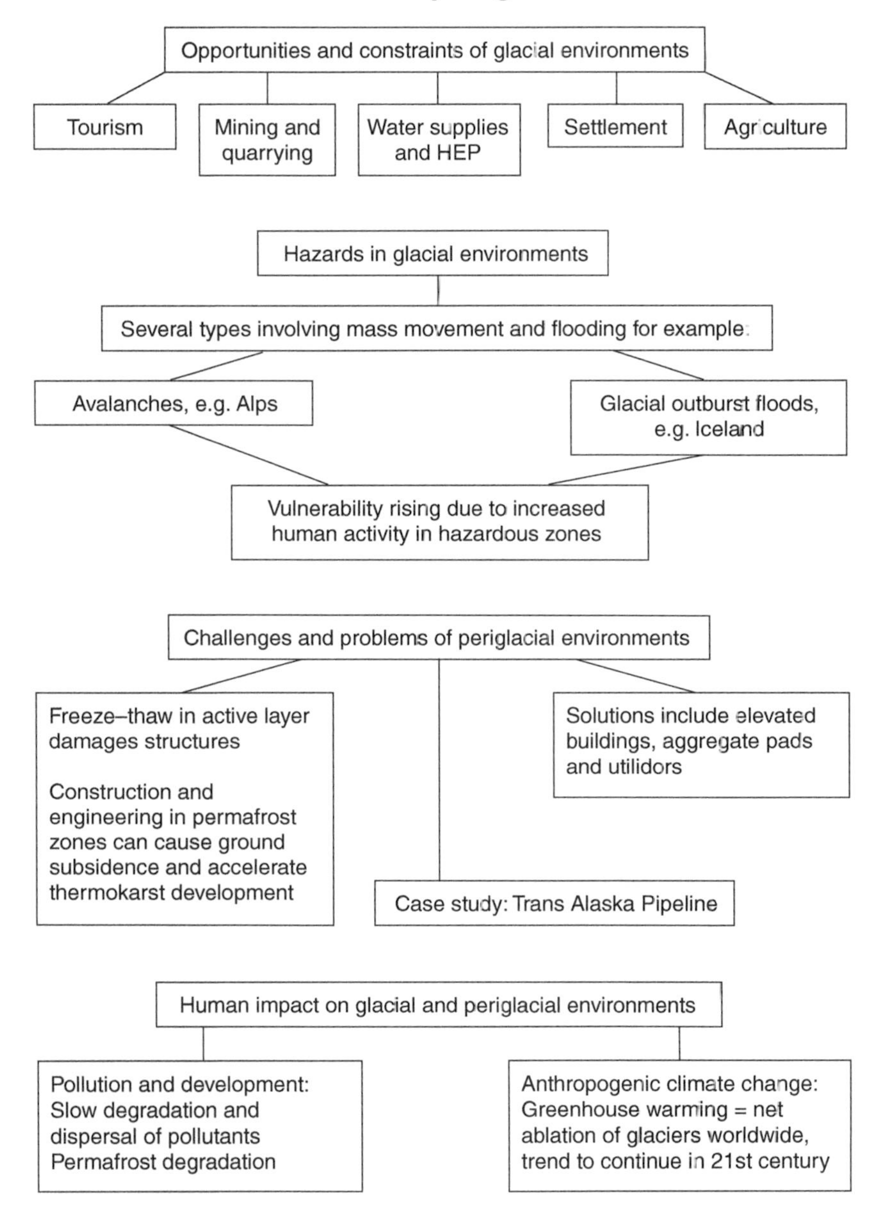

## Questions

1. With reference to specific areas, describe how human activities are influenced by the glacial environment.
2. Glaciated areas provide many opportunities for sustainable economic development. Discuss this statement.
3. Discuss how human vulnerability to hazards of glacial environments can be reduced.
4. **a)** Why do human activities cause degradation of permafrost?
   **b)** Describe the problems caused by permafrost degradation.
   **c)** Using examples, discuss the solutions available to protect permafrost.
5. **a)** Explain why pollution and development can cause severe and long-lasting problems in periglacial environments.
   **b)** Describe and explain the effects of anthropogenic climate change on the glacial environment over the short term.
   **c)** Why is it difficult to predict changes in the distribution of glacial and periglacial environments in the long term?

# Bibliography


Anderson, D., 2000, Abrupt climatic change. *Geography Review* 13, 2–6.

Ballantyne, C.K., 2001, The sorted stone stripes of Tinto Hill. *Scottish Geographical Journal* 117, 313–24.

Ballantyne, C.K. and Harris, C., 1994, *The Periglaciation of Great Britain* (Cambridge: Cambridge University Press).

Benn, D.I. and Evans, D.J.A., 1998, *Glaciers and Glaciation* (London: Arnold).

Bennett, M.R. and Glasser, N.F., 1996, *Glacial Geology: Ice Sheets and Landforms* (Chichester: Wiley).

Boardman, J. and Walden, J., (eds.), 1994, *The Quaternary of Cumbria: Field Guide* (Oxford: Quaternary Research Association).

Briggs, D., Smithson, P., Addison, K. and Atkinson, K., 1997, *Fundamentals of the Physical Environment,* 2nd edition (London: Routledge).

Clowes, A. and Comfort, P., 1987, *Process and Landform: An Outline of Contemporary Geomorphology* (Harlow: Oliver and Boyd).

Collard, R., 1988, *The Physical Geography of Landscape* (London: Unwin).

Evans, D.J.A. and Hansom, J.D., 1996, The Edinburgh Castle crag-and-tail. *Scottish Geographical Magazine* 112, 129–31.

Evans, I.S., 1997, Cirques and moraines of the Helvellyn Range, Cumbria: Grisedale and Ullswater. In Boardman, J., (ed), *Geomorphology of the Lake District: A Field Guide,* pp. 63–88 (Oxford: BGRG).

Ferguson, S.A., 1992, *Glaciers of North America: A Field Guide* (Golden, Colorado: Fulcrum).

Fishpool, I., 1996, Glaciation and deglaciation: The Columbia icefield and Athabasca glacier. *Geography Review* 9, 11–15.

Foster, S., 2003, Landforms on Dartmoor: Part 2. Frozen in the landscape. *Geography Review* 16, 38–41.

French, H.M., 1976, *The Periglacial Environment* (London: Longman).

Goudie, A., 1990, *The Landforms of England and Wales* (Oxford: Blackwell).

Goudie, A., 1992, *Environmental Change,* 3rd edition (Oxford: Clarendon).

Goudie, A. and Gardner, R., 1992, *Discovering Landscape in England and Wales* (London: Chapman & Hall)

Gribbin, J. and Gribbin, M., 2001, *Ice Age* (London: Penguin).

Guinness, P. and Nagle, G., 2002, *Advanced Geography: Concepts and Cases* (London: Hodder & Stoughton)

Lowe, J.J. and Walker, M.J.C., 1997, *Reconstructing Quaternary Environments,* 2nd edition (Harlow: Longman).

Lynas, M., 2001, Melting points. *Geographical* 73, 42–8.

Money, D.C., 1980, *Polar Ice and Periglacial Lands* (London: Evans Brothers).

Moore, P.D., Chaloner, B. and Stott, P., 1996, *Global Environmental Change* (Oxford: Blackwell).

Nuttall, M. and Callaghan, T.V., (eds), 2000, *The Arctic: Environment, People, Policy* (Amsterdam: Harwood).

Pearce, E.A. and Smith, C.G., 1993, *The World Weather Guide,* 3rd edition (Oxford, Helicon).

Smith, R.B. and Siegel, L.J., 2000, *Windows into the Earth: The Geologic Story of Yellowstone and Grand Teton National Parks* (Oxford: Oxford University Press).
Smithson, P., Addison, K. and Atkinson, K., 2002, *Fundamentals of the Physical Environment*, 3rd edition (London: Routledge).
Sugden, D.E. and John, B.S., 1976, *Glaciers and Landscape* (London: Edward Arnold).
Summerfield, M.A., 1991, *Global Geomorphology* (Harlow: Longman).
Wilson, R.C.L., Drury, S.A. and Chapman, J.L., 2000, *The Great Ice Age* (London: Routledge and The Open University).
Wilson, S.B. and Evans, D.J.A., 2001, Coire a' Cheud-chnoic, the 'hummocky moraine' of Glen Torridon. *Scottish Geographical Journal* 116, 149–58.

## Websites

Intergovernmental Panel on Climate Change: http://www.ipcc.ch/
National Snow and Ice Data Center (NSIDC): http://nsidc.org/index.html
Trans Alaska Pipeline: http://www.alyeska-pipe.com/
United States Geological Survey (USGS) on glaciers and glacial hazards: http://vulcan.wr.usgs.gov/Glossary/Glaciers/
World Glacier Monitoring Service: http://www.geo.unizh.ch/wgms/

# Index